T0075264

About Island Press

Since 1984, the nonprofit organization Island Press has been stimulating, shaping, and communicating ideas that are essential for solving environmental problems worldwide. With more than 1,000 titles in print and some 30 new releases each year, we are the nation's leading publisher on environmental issues. We identify innovative thinkers and emerging trends in the environmental field. We work with world-renowned experts and authors to develop cross-disciplinary solutions to environmental challenges.

Island Press designs and executes educational campaigns, in conjunction with our authors, to communicate their critical messages in print, in person, and online using the latest technologies, innovative programs, and the media. Our goal is to reach targeted audiences—scientists, policy makers, environmental advocates, urban planners, the media, and concerned citizens—with information that can be used to create the framework for long-term ecological health and human well-being.

Island Press gratefully acknowledges major support from The Bobolink Foundation, Caldera Foundation, The Curtis and Edith Munson Foundation, The Forrest C. and Frances H. Lattner Foundation, The JPB Foundation, The Kresge Foundation, The Summit Charitable Foundation, Inc., and many other generous organizations and individuals.

The opinions expressed in this book are those of the author(s) and do not necessarily reflect the views of our supporters.

Humanity's
Moment

Humanity's Moment

A Climate Scientist's
Case for Hope

Joëlle Gergis

◯ **ISLAND**PRESS | Washington | Covelo

Library of Congress Control Number: 2022942434

All Island Press books are printed on environmentally responsible materials.

Manufactured in the United States of America
10 9 8 7 6 5 4 3 2 1

Conversion Chart
1° Celsius = about 34° Fahrenheit (Multiply degrees Celsius by 1.8 and add 32.)
1 kilometer = about 0.62 mile
1 meter = about 1.10 yards
1 kilogram = about 2.2 pounds
1 Australian Dollar = about 0.78 US Dollar
1 tonne = 2204.6 pounds

Keywords: activism, atmosphere, Australia, biodiversity, carbon cycle, clean energy, climate anxiety, climate change, climate policies, climate science, CO2, extreme weather, feedback loops, greenhouse effect, ice sheets, Intergovernmental Panel on Climate Change, IPCC, nitrogen, Paris Agreement, sea level rise, social movements, tipping points

For Josh; your light keeps me here ...

Life on the frontline

IT'S PITCH BLACK AS I slip out of bed, trying not to wake my husband. It's 5:15 a.m. on a Saturday in the dead of winter 2020; the last thing I feel like doing is leaving my downy cocoon to talk about our destabilizing world.

As one of the dozen or so Australian lead authors involved in the United Nations' Intergovernmental Panel on Climate Change (IPCC) *Sixth Assessment Report*, it's my job to review thousands of peer-reviewed scientific studies and distill their key findings. Our task in Working Group 1 is to provide the scientific foundation for understanding the risk of human-induced climate change, its potential impacts, and options for adapting to and avoiding dangerous levels of climate change. The cycle typically takes around six years to complete, from initial scoping to final government approval. Despite the outbreak of a deadly pandemic, work continues on, as the stakes are now so high. Our assessment forms the technical foundation for trying to achieve the Paris Agreement targets of keeping global warming to well below 2°C above pre-industrial levels, and as close to 1.5°C as possible. It's the information that helps us figure out how to keep our planetary conditions safe for humanity.

The first volume of this global climate assessment report is due out in mid-2021, and this morning's meeting is one of many since the process started back in June 2018. Our second draft has just come back from government and expert review; we now have 51,387 technical comments to address ahead of the Final Government Draft that will go to the UN for approval. Today's task is to come up with a strategy for responding to each comment assigned to our chapter and revising our text to meet our deadline.

Before embarking on the IPCC process, I had managed to remain emotionally detached from the work that I do. I could focus on my research, reporting findings, and not allow their implications to sink in too deeply. I was okay as long as I didn't look at images of scorched animals, distressed farmers and ravaged landscapes long enough to feel the sting of it. My work felt clean, clinical, safely partitioned from my emotions. But being involved in a UN process reconfigures your worldview; it forces you to zoom out and take in the world as a whole. My chapter team alone spans scientists from Colombia to France, Russia to Cameroon, Israel to India. I'm lucky to form part of our team's trans-Tasman contingent; the Australia–New Zealand alliance, solid as always.

As I roll out of bed on this dark winter's morning, my nerves are shot: it's been a hell of a year. But if I don't get up, I'll miss the opportunity to represent Australia's scientific community in this global process. At the rate we're going, by the time the next IPCC report comes out – likely around 2030 – the world will have blown the carbon budget required to achieve the Paris Agreement targets. In my darker moments, I fear that we may have already crossed an invisible threshold, pushing the planetary system past the point of no return. So as much as I need the rest, it's time to haul myself out of bed and take one for the team.

* * *

As I jot down my task list in the notepad on my desk, I listen to the pre-dawn chorus – first the kookaburras, then the lorikeets, followed by the plaintive currawongs. By the time the meeting ends, my usual alarm clock – a huge flock of noisy corellas – has screeched overhead, herald-ing the start of a new day. Although I sometimes feel aggrieved by the terrible time zones of these meetings, more often than not I feel lucky to live in a landscape that still feels so alive, still part of a thriving eco-system. Over 200 bird species make this pocket of northern New South Wales, Australia, their home. At these ungodly hours, nature thrives, reminding me that so much hangs in the balance in a country like mine. We are one of the planet's most biodiverse, with more unique plant and animal species than anywhere else on Earth. Like Brazil, Papua New Guinea and Madagascar, Australia is a global biodiversity hotspot. As David Attenborough recently reminded us: "You are the keepers of an extraordinary section of the surface of this planet . . . what you say, what you do, really, really matters."

United Nations member countries put forward nominations for 911 regional experts to compile the IPCC's *Sixth Assessment Report*, all of us assessed by our contribution to published scientific research. From these nominations, 234 scientists from sixty-six countries were selected to serve as lead authors on the Working Group 1 report to pro-vide the necessary expertise to conduct the assessment across a range of disciplines. Just over a quarter of us are women, with the southwest Pacific – which includes Australia – accounting for only 9 percent of the voices at the table.

But no matter where we are from, all of us volunteer our time, work-ing thousands of unpaid hours over the course of three relentlessly intense years, drafting technical summaries of complex topics including the causes of chronic drought, modeling sea level rise and changes in tropical monsoons. Our work goes out for two rounds of expert and gov-ernment review – alongside countless internal checks – after which we face the epic task of responding to tens of thousands of reviewer comments and attending meetings across several time zones to make sure the revisions get done. In the end, our group will respond to 78,007 technical queries throughout the review process – all made publicly available in a database alongside the final report for complete transparency.

All of this happens against the background of our day jobs – generally, positions as research scientists and university lecturers in world-leading institutions – and the immovable challenges of domestic life. In Australia, the only funding is a small federal government travel allowance to cover economy flights, standard accommodation and basic meals associated with attending four compulsory in-person lead author meetings.

The first three involve long-haul flights to China, Canada and France for incredibly intense five-day meetings. The last session done, I catch the first evening flight to begin the long journey home, often through multiple international airports, while trying to make a dent in the mountain of non-IPCC email that accumulated during the week.

On the way home from our third meeting in France, a mix of anxiety and fatigue has my mind racing. I can't sleep; I'm too overstimulated. I pull out my journal to capture the rush of ideas, page after page. The last thing I write that night still haunts me:

> It's extraordinary to realize that we are witnessing the great unraveling; the beginning of the end of things. I honestly never thought I'd live to see the start of what sometimes feels like the apocalypse. The Earth is struggling to maintain its equilibrium. It's possible that we are now seeing a cascade of tipping points lurching into action as the momentum of instability takes hold and things start to come apart. I honestly don't know what the future will bring.

When complete exhaustion sets in, I stare out the window, weeping at the enormity of the challenge we face. It's a battle between despair and belief in the power of true global citizenry that I've just experienced.

*   *   *

When COVID-19 strikes, our fourth in-person meeting – slated for June 2020 in Chile – is canceled and rapidly adapted into virtual form. Part of me welcomes the opportunity not to put my body through the punishment. But the gathering is replaced by more than a dozen online meetings over several months, often involving brutal time differences – all while I am teaching a new climatology course for 120 students that is suddenly forced to shift online as my university campus shuts down.

During one meeting, Cyclone Nisarga slams into western India – the strongest tropical cyclone to strike the Indian state of Maharashtra since 1891. Despite this, an author based there manages to log in, apologizing for being late:

> Hello everyone – sorry for the delay. The internet and electricity was totally down due to a cyclone that passed over Pune just now. I am trying to reconnect using mobile hot spot. The connection is weak and intermittent.

In light of the lived immediacy of these struggles, my own challenges feel insignificant.

The dedication shown by my fellow scientists is truly the stuff of legend. During the Australian summer of 2020–2021, as most people enjoy their Christmas holiday, climate scientists across the world work around the clock to complete this monumental climate change assessment during a global pandemic. For those of us from the Southern Hemisphere, it's the third year in a row we've worked through our summer break. The fatigue that sets in is bone-deep. Sometimes I doze off in the middle of the day, waking fitfully from vivid dreams that leave me feeling trapped in a perpetual state of jet lag. Often, I slip into bed before sunset, wrecked, unable to do anything other than sleep.

Grueling workloads aside, we all enjoy the cultural exchange that comes with being part of a team of scientists gathered from so many countries, working to compile the most comprehensive global stock-take of climate change humanly possible. My favorite lunch buddy is the lone delegate from Iceland with a name so unpronounceable that she graciously offers up a nickname. We communicate warmly with our eyes when words fail us. We figure out that we live about as far away from each other as is physically possible. Our mealtime chats revolve around melting glaciers, the complexities of modeling sea level rise and if Australia really has a proper winter.

Back in formal meetings, we spend hours listening to how the climate crisis is escalating all over the world. Warming deep in our oceans. Melting glaciers in densely populated lands. Rainfall patterns drifting away from continents, slipping towards the poles. We talk non-stop about real-time examples of accelerating warming already observed in our unique parts of the world. We worry about how quickly things are playing out, orders of magnitude faster than the natural processes of geologic and evolutionary time. The more I hear, the more I realize that the situation is far worse than most people can imagine. In truth, it's also hard for me to fathom that our generation is likely to witness the destabilization of the Earth's climate; that we will be the last to see the world as it is today.

That's what really keeps me up at night. I wonder if we may have already pushed the planetary system too far, unleashing a cascade of irreversible changes that have built such momentum we can now only watch as they unfold. All the latest models assessed in the IPCC report see us sail through the Paris Agreement's 1.5°C target in the early 2030s; that's now just ten years away.

*   *   *

It was impossible to imagine what would unfold after our third lead author meeting in France in August 2019. After sweltering through an extreme European summer heatwave, I returned home to find much of Australia's eastern seaboard engulfed in an unprecedented bushfire crisis, scarcely a week out of winter. The catastrophic conditions were the result of the hottest and driest year on record – the first time both national records were broken in the same year. Rainfall was a staggering 40 percent below average across the entire country, with temperatures running 1.5°C warmer than the historical mean.

Close to 150 fires were raging as 98 percent of New South Wales and 65 percent of Queensland baked through one of the most punishing droughts in Australian history. Fire maps of the east coast lit up like a grotesque Christmas tree. Many were burning out of control as the relentlessly gusty winds grounded the firefighting aircraft needed to try and contain the blazes. Some regional towns that were already trucking in water to cope with the relentless drought now had fires bearing down on them.

For me, the worst were the wildfires in the World Heritage–listed rainforests of Lamington National Park in the Gold Coast hinterland of Queensland and Nightcap National Park on the north coast of New South Wales. These regions are part of the Gondwana Rainforests of Australia, an area that contains the largest remaining stands of subtropical rainforest in the world, as well as the most significant areas of warm temperate rainforest in the country. These usually moss-drenched forests are packed with the oldest elements of the world's ferns and primitive plant families, dating back to the Jurassic era, some 200 million years ago. They are indescribably precious. Although these remarkable ecosystems have clung on since the age of the dinosaurs, searing heat and bone-dry conditions saw these usually lush forests turn into fuel.

These areas are close to my heart. The rainforests of northern New South Wales are where my husband spent much of his time growing up.

Since we met over twenty years ago, he's taken me to explore remnant patches of these primordial places he played in as a boy. Whenever we get a chance, we head for these relic forests, a reminder of a time when the Earth was still young. The blockades to protect the Terania Creek basin in Nightcap National Park from logging in 1979 were the first of their kind in the Western world. They went on to inspire the protests that saved the Franklin River in the Tasmanian wilderness during the early 1980s, which in turn led to the development of the first Green political party in the world. These extraordinary forests deep in the Australian rainforest helped birth the global environmental movement.

It's where I go when I am burnt out and heartsick. They help me remember that the Earth is still alive, that there are areas worth saving. Perhaps more importantly, they remind me that these places still exist because humans cared enough to protect them. They reconnect me with what feels like the very best of humanity; something deeply profound, intergenerational.

By the time Australia's Black Summer finally came to its horrific end, fire had torn through Terania Creek, along with 53 percent of the last of these ancient Gondwana rainforests. When that news came through, I sat at my desk sobbing, knowing that these areas are likely to be lost forever. Something inside me broke.

That horrendous summer saw more than 3 billion animals incinerated or displaced by an unimaginable path of destruction, ripping through globally significant biodiversity hotspots across the country. Around 23 percent of Australia's forests burnt in a single bushfire season. The world saw images of terrified animals fleeing with their fur on fire, their bodies turned to ash. Those that survived faced starvation in the charred remains of their obliterated homes. The koala, Australia's most emblematic species, lost so much of its habitat that it now faces extinction in the country's most populous state of New South Wales as early as 2050. In February 2022, koalas in eastern Australia were officially added to the endangered species list, something I never imagined I'd witness in my lifetime.

But what really scares me is what Australia's Black Summer says about things yet to come. In the aftermath, scientists analyzed the conditions observed during the 2019–2020 fire season, concluding that "under a scenario where emissions continue to grow, such a year would be average by 2040 and exceptionally cool by 2060." That is, the most extreme statistical outliers in today's climate will become average conditions in just twenty years. Soon, searing temperatures over 50°C will become a regular feature of summer in Sydney and Melbourne, where around 40 percent of the nation's population lives.

Australians won't be the only ones forced to endure searing heat: beyond 2°C of global warming, maximum summer temperatures will reach 50°C across all continents, with projected temperatures above 60°C in hotspots like Pakistan, Iraq and Saudi Arabia. By 2050, Madrid's climate will resemble Marrakech's climate today, London will feel like Barcelona, and Seattle more like San Francisco.

The latest climate models show that under a very high emissions pathway, global average temperatures warm as much as 3.3–5.7°C above pre-industrial levels by the end of this century, with a central estimate of 4.4°C. But because the planet is mostly covered in ocean that absorbs excess heat, thinking about global averages doesn't give you a true sense of the warming that will occur over land, where people actually live. Most continental areas will warm well above the global mean. For example, under a worst-case scenario, average land temperatures along the eastern seaboard of North America are projected to rise by 6.5°C above pre-industrial levels by the end of the century, with all models simulating a range of 4.7–8.7°C. In northern European regions like the United Kingdom, average temperatures warm by 6.3°C, with projected temperatures ranging between 4.3 and 9.0°C by 2100. Even under an optimistic intermediate emissions scenario, average warming in both these areas, which house the great cities of New York and London, is 4.0°C by the end of the century. And these are just averages: extreme temperatures are far more pronounced, often changing twice as fast as the long-term mean.

Under a fossil fuel–intensive scenario, land areas of Australia are projected to warm between 4.0 and 7.0°C above pre-industrial levels by 2100, with a central estimate of 5.3°C. This level of heat will render large parts of the country uninhabitable. Such high levels of warming will profoundly alter not only Australia, but all life on Earth. Although Australians wear our badge of resilience with a hefty dose of national pride, scientists understand that some things in life, once gone, can never be replaced. There is only so much the system can take. If the new models are right, there is no way we can adapt to such catastrophic levels of warming.

*   *   *

As someone on the frontline of research on the climate crisis, I try to help people make sense of the latest results coming out of the scientific community. But when the new climate projections were published, I found it impossible to focus.

So, I wrote my way through it, and published a long-form essay in July 2020. I'd been afraid to publish such a personal piece, fearing my colleagues would think less of me for sharing my emotional response to our work. But I took heart from a quote by Rachel Carson, ecologist and author of the seminal book *Silent Spring*, first published in 1962:

> It is not half so important to know as to feel . . . once the emotions have been aroused – a sense of the beautiful, the excitement of the new and unknown, a feeling of sympathy, pity, admiration or love – then we wish for knowledge about the object of our emotional response. Once found, it has lasting meaning.

In other words, there is power and wisdom in our emotional response to our world. Until we are prepared to be moved by the profoundly tragic ways we treat the planet – and each other – our behavior will never change.

As scientists, we are often quick to reach for more facts rather than grapple with the complexity of our emotions. Science can seem cold and complicated; scientists, detached and dull. But as the long history of humanity's inability to respond to the climate crisis has shown us, processing information on a purely intellectual level just isn't enough. I've come to realize that no amount of extra data or technical graphs is going to help people actually *feel* the grief of what we are facing. Although it's not the accepted practice in our field, I felt compelled to share the immense loss I felt – not just as a scientist, but as a human being. Perhaps if I am honest about my own emotional response to our work, it might help others feel something too. In the words of American civil-rights activist Rosa Parks: "Knowing what must be done does away with fear."

When my article was published, I received an email from an IPCC colleague in a far-flung corner of the world:

> I've been deeply depressed since the meeting in Singapore . . . I almost lost my position here at the university because I could not care less about work knowing that we seem to be doomed. I just wanted to sleep and do nothing . . . I then realized I was depressed and . . . on a kind of autopilot, just doing the mere essential (of course that also included fulfilling the IPCC deadline in January) but everything looks black and void . . . and then I read your article, and I realize that I am not the only one in despair given how little time we have to make radical changes, and realizing that people are not keen at all to do so. I still worry, it's still on my mind most of the time, but I can function somewhat normally now. I wonder how many of us feel like that, and are able to actually say it?

It's a long-held myth that a credible scientist should be devoid of human emotion, presenting our work rationally, without commentary. Perhaps it's part of the reason why the public discussion around climate change has been so dominated by the political right across the developed world – the environmental, social and cultural costs of capitalism have been dismissed in the name of economic progress for far, far too long. Given that humanity is now facing an existential threat of planetary proportions, and scientists are the people who really know exactly what's at stake, shouldn't that logically include acknowledging our sense of despair, anger, grief and frustration? Why are medical doctors praised for a good bedside manner, while climate scientists are dismissed as "alarmist" if we express our deep concern about the state of the world? Would anyone ridicule an intensive care nurse for feeling distressed if someone in their care died on their watch? Is it possible to witness the death of the Great Barrier Reef – the largest living organism on the planet – and not feel wild with desperation at the thought of it all?

*   *   *

My involvement in the IPCC process has been life changing. It's been overwhelming to get a complete sense of how the planet's climate is changing – on all levels, at all timescales, all over the world. The scientists I work with are aware that we might only know we've crossed critical thresholds in hindsight. The warning signs of tipping points are all there: the rapid melting of polar regions, the freshening of the North Atlantic Ocean, and widespread fires in the Amazon. Now, more than ever, we need more scientists on the frontline to sound the alarm, no matter how uncomfortable we may feel.

We know from the geologic record that 1.5 to 2°C of warming is enough to seriously reconfigure the Earth's climate. In the past, such changes triggered substantial long-term melting in Greenland and Antarctica, unleashing 6–13 meters of global sea level rise lasting thousands of years. With the 1.2°C of global warming we've experienced so far, an alarming proportion of the world's coral reefs have already experienced large-scale die-off. Between 2016 and 2017, the Great Barrier Reef lost approximately 50 percent of its shallow water corals following unprecedented back-to-back mass bleaching events. As there are long gaps in reef-wide monitoring, it is still unknown exactly how much more died during the mass bleaching that struck the reef again in March 2020, the most widespread event ever recorded in the region. And then, to the horror of the scientific community, in March 2022 yet another mass bleaching engulfed the beleaguered reef – the fourth since 2016. It is clear that the largest living organism on the planet is in terminal decline. It's truly the stuff of nightmares.

Even if it were still geophysically possible to achieve the most ambitious goal of limiting warming to 1.5°C, we will still see the destruction of 70–90 percent of coral reefs that exist today. With 2°C of warming, 99 percent of tropical coral reefs disappear. An entire component of the Earth's biosphere – our planetary life-support system – will be destroyed. The domino effect on the 25 percent of all marine life that depends on these areas will be profound and immeasurable.

Right now, *current policies in place today* will lead to 1.9–3.7°C of warming by the end of the century, with a best estimate of 2.6°C. This represents a catastrophic overshooting of the Paris Agreement targets, which were specifically developed to avoid "dangerous anthropogenic interference with the climate system." If countries fully implement their long-term net-zero emissions targets, this best-case scenario could see global warming stabilize between 1.4 and 2.8°C by 2100, with 2°C considered most likely. The problem is there are no guarantees that countries will honor their commitments, as only fourteen of the 196 parties have formalized net-zero targets into legislation, meaning the majority of pledges are still not legally enforceable. To have a chance of limiting warming to 1.5°C by 2100, global emissions need to halve by 2030. This means the world needs to more than *double* its current emissions-reduction pledges to restrict warming to 1.5°C.

We have a hell of a job ahead of us.

*   *   *

Sometimes I'm unsure of how best to live my life in the face of the catastrophe that is currently unfolding. Between IPCC, research, teaching, and grieving for our planet, sometimes I feel I have nothing more to give.

As coronavirus lockdowns lift and national parks reopen, the first thing we do is pack the car and head for the rainforest. Driving up the windy access road of the Border Ranges National Park feels like going to check in with family after a disaster – we're afraid of what we might find. As we stand at our favorite lookout, I find it hard to see through tears. My husband pulls me in close and whispers, "It's still here. It's still here." This immense valley drenched in brilliant green; the rainforest we love so much.

These magnificent forests have survived for millions of years. My hope is that they can hang on, that the cavalry is on its way. As a climate scientist, I am doing everything I possibly can to respond to the distress signals from our natural world. If I live to look back at this troubled time, I want to say that I did all that I could, that I was on the right side of history.

*   *   *

After reading David Attenborough's *A Life on Our Planet*, his moving "witness statement" of observing the decline of the natural world for nearly seventy years as a documentary filmmaker, I realized that climate scientists also have an important story to tell. This extraordinary 96-year-old, a man who has probably seen more of the planet than anyone who has ever lived, someone who has inspired generations of people to connect with the wonders of nature, is using the time he has left to bring the urgency of the climate emergency to the public. He cannot be silent knowing what he knows is at stake.

It's hard to care about things you cannot see.

So I'm writing this book to provide you with an insider's account of the latest research and what it is like being a scientist involved in the IPCC; the UN's peak body responsible for assessing global climate change. When IPCC reports are released, they are often filtered through the media and a range of other non-expert commentators. We might get a short quote in a news article or, if we are lucky, some airtime on the radio or television, but scientists very rarely get a chance to tell our stories, in our own words. Some of us try to write technical explainers or media releases to share our latest research with the public as quickly as possible, but usually we are so tied up doing the actual science, and we don't often have the training or the time to communicate the implications of our research. Fewer still have the courage to publicly confess the emotional toll our work is increasingly having on our lives.

But after contributing to the latest IPCC assessment report, it became alarmingly clear that people outside of the scientific community honestly don't appreciate the scale of the crisis that is currently unfolding. Our science is not translated accurately or quickly enough to reflect just how serious the situation has become. There are many reasons for this, but I'll outline just a few. Because IPCC assessments must be as objective and thorough as possible, our key messages are often buried deep in technical detail that we agonize over endlessly (trust me, I never want to hear about the role of aerosols in cloud microphysics ever again). As a result, it's often hard for people to grasp the true extent of the problem, or they end up with a skewed version of reality when our results are inadvertently misinterpreted or actively misconstrued.

Very often the IPCC's most important messages are lost in translation. Our science can be dense and is written with so much nuance that it's sometimes hard to understand what we are trying to say. This is because the science itself is incredibly complex, and we need to phrase our findings using "uncertainty language" that must remain solidly anchored in the evidence base. Each sentence that appears in the final report is made up of qualitative "confidence statements" that take into account the type, quality, amount and consistency of evidence, and the overall agreement, or consensus, that can be reached based on the best available science. So if a research area is still emerging as new observations or theoretical developments come to hand, an IPCC assessment may report a statement as "low confidence" even if there is a high level of agreement based on the available evidence to date, simply because there may only be a few studies published on a particular topic. It doesn't mean that the underlying science isn't solid, it just means that it's an active research area on the cutting edge of the field.

To further complicate things, alongside our confidence ratings, the IPCC also provides "likelihood statements," which are quantitative assessments based on statistical analyzes or model results. Essentially this process assigns a probability of a given statement being true. For example, the IPCC considers something to be "unlikely" if it corresponds to a probability range of 0–33 percent, calculated from the best available data. For a statement to be considered "very likely" we are looking for a probability of 90–100 percent – something that sits well outside the realm of being a statistical fluke. And because we are a group of obsessive perfectionists, of course there are ten tiers of likelihood statements to cover all bases.

It's important to understand that low-likelihood scenarios still pose a very real threat – it's just that the risks can't be predicted clearly based on currently available evidence. These scenarios may include well-understood physical processes, like a large volcanic eruption or ocean dynamics, but might be statistically rare or unprecedented in short observational records, or difficult to model. The IPCC classifies these as "low-likelihood, high-impact" scenarios based on what we understand about the science *right now*. But if things play out consistently with what we know so far about, for example, the speed of abrupt climate change seen in geological records, the impacts on society could be catastrophic. For the first time, this IPCC assessment goes to great lengths to explain that these high-risk scenarios cannot be ruled out. It clearly states that there is an increased chance of substantially larger global or regional changes than the "very likely" range reported for future projections, particularly under higher levels of global warming.

As you can see, understanding climate change is a little bit like trying to complete a jigsaw puzzle without all the pieces available. Everyone is working incredibly hard to place the pieces we do have, but science takes time. Its measured pace is in direct conflict with the relentless speed of the news cycle. This means that the loudest voices – which are not necessarily the most informed, or even sane – often dominate public commentary. Our work becomes politicized by those not constrained by the professional ethics and rigor of our discipline, resulting in speculative discussions on complex topics that many scientists aren't comfortable covering. Hence our use of carefully crafted uncertainty language, which is at odds with the fast and loose approach others are willing to take. Sensational clickbait always seems to win.

We live in an era when reliable information matters more than ever. As I see it, our messages are ignored by the public not because people don't care, but because most people don't have a science background, or find it challenging to stay engaged in a technical debate that feels so far removed from the reality of their lives. The discussion around climate change has become so divisive that it's hard for most people to feel like they can be part of the conversation. It's hard to know what's true or who to trust, and eventually it all becomes too overwhelming, so we switch off. Meanwhile, all over the world, elected politicians are making fateful decisions right now that will shape the entire course of humanity's future. In the end, even those with political and economic power aren't even aware of how bad things have become. When we tune out, we squander the most powerful thing we have to influence the system – our vote.

During my time working on the *Sixth Assessment Report*, it dawned on me that this IPCC assessment is probably the scientific community's last chance to really make a difference. If our work doesn't convince this generation of political leaders that we must stabilize the Earth's climate immediately, we will lock in an irreversibly apocalyptic future. I realized that the most important thing I can do right now is not write another research paper, apply for more funding or teach another climatology course – the most important thing I can do right now is share everything I've learned as far and as widely as I possibly can. I've distilled our key findings down as simply as possible so you can clearly see that what we fear is now on the horizon. I want you to understand that the extreme conditions you have been experiencing in your part of the world are not only happening where you live, but are part of a global trend that has experts very, very worried.

Climate change is real and it is here, and it's not going away. We need your help.

Just like a doctor diagnosing a critically ill patient, as a climate scientist I face the terrible task of being the bearer of bad news. It is akin to asking each person to sit with the horror and grief of the prospect of losing the very life force that miraculously sustains us all. I need to take you by the hand and gently ask you to stay with the gravity of what's at stake and what it means for your future. But just like a serious health condition, if symptoms are caught early enough, appropriate treatment can avoid a condition progressing into a terminal situation. If we intervene before it is too late, there are things that can be done to lessen the impacts. Tumors can be removed, lifestyles can be changed, lives can be saved.

When it comes to the planet, we need to understand that what we do collectively *right now* will shape the future course of humanity. You only have to hear a statistic like 90 percent of seabirds alive today contain plastic – a by-product of the fossil fuel industry – in their guts to know that humanity has lost the way. It's fair to say that we have completely overrun the Earth; it desperately needs our active protection to stay alive. We have urgent choices to make about how much will be lost to future generations. We must choose what we are willing to save.

*   *   *

I'm guessing you've picked up this book because you already know that our world is changing. Maybe you are trying to come to terms with what climate change means for your life right now. Perhaps you've lived through a traumatic event that impacted you in a personal way: your house burnt down, your family was displaced by floodwaters, or a place you love is disappearing before your eyes. Maybe you grew up loving summer, but these days, the heat outside scares you – you know a shift in the wind can destroy everything you hold dear. The world needs your resilience and insight to help guide us through loss.

You might be a parent and want to understand the future your kids are now facing. Maybe you, too, find yourself weeping watching David Attenborough documentaries because you realize the little ones in your life will never experience the world the way you did when you were growing up. You want to know what you can do to protect their future, to try and save what's left. You know that generational change starts at home. We need your love and compassion to remind us why we cannot fail.

Or perhaps you are a young person who already understands how serious this is. You've taken to the streets in solidarity with millions from your generation, fueling the social movement sweeping the world. You want to connect with others who care about the future of our planet, and arm yourself with enough accurate science to pass on wherever you can. You probably feel angry with the people who let things get so bad; trust me, I do too. The world needs your spirit and drive to help redirect the course of our future.

Maybe you've been on board for decades – you might be a business leader, an environmentalist, a community activist, a politician or a media professional – and you just want someone to clearly step you through the highlights of the latest IPCC report. You are wondering if this will be the moment when things really change, when the momentum you've been building for so long will finally tip the system. Maybe you need a reminder that all the transformational struggles throughout human history have passed through the gates of despair on the way to victory. That all revolutions seem impossible until they become inevitable. We need your wisdom and inspiration to help us hold on.

Or maybe you are a creative, someone who feels things so deeply that it hurts. You care but are overwhelmed by how hard everything feels sometimes. You don't know what you can do as a musician, an artist or a writer to stop climate change, but you know how to express yourself in ways that help other people feel things too. You remind the rest of us that there is still so much beauty in the world, that we must allow ourselves to be moved enough to save it. We need your sensitivity and imagination to help us reclaim our soul. Promise you will stay with me until the end.

If you are a scientist, a researcher or a teacher trying to make sense of how you've been feeling about work lately, it might seem like you are trying to hold back the tide in your personal and professional life. Some days you sense you're fighting a losing battle. Other times, you know you are changing the world, one careful step at a time. My hope is that you find solace in these pages – a reminder that the work you do is not invisible. What you care about really, really matters, even if others can't see that yet. Your dedication and vision will be valued by generations to come.

No matter who you are, above all, you are ready to connect your head with your heart. You are devastated by what you see and can no longer turn away. You want to do what you can, where you can, with the time you have left. This book is an invitation to reclaim our shared humanity at this transformative time in history, wherever you are in the world. You want to survive the journey through the heartland of your grief and create a meaningful life on the other side. You want to be a part of the group of people who cared enough to try.

Wherever you are on this path, I want to offer you my company, to let you know that you are not alone. I am sharing my personal response to facing this unimaginable dilemma, and how I've tried to navigate my own despair. I want you to know that the fear you feel is rational. So is your pain. I understand the science can often be complicated and intimidating; I'll do my best to make it digestible for you. I will shine a light into my world and provide insights I've gained from being among a group of the world's leading climate scientists trying to avert disaster at this critical moment in human history.

I definitely don't have all of the answers – I'm navigating this radical transformation of the world just like you – but I'll draw together everything I've found helpful to contribute to your conversations. You'll see that the solutions we need to live sustainably on our planet already exist right now – we just need the social movement and the political will to create a better world. I hope that this book will transform your feelings of grief and anger into action and a genuine sense of hope. I want you to see how you can contribute your talents to help create our new world, no matter your circumstances, or how small your efforts might feel. All any of us can ever do is show up and do what we can with what we've got.

Like all great social movements, everyone, everywhere, is needed. It's time for our business leaders, our musicians, our filmmakers, our politicians, our teachers, our families – all fellow humans – to step up and help reimagine a future where beauty and heartbreak strike a tolerable balance. It's time to get behind leaders who will act with courage and vision that future generations will be proud of. There have always been people across the ages who have risen to face the great challenges of their time and succeeded against all odds. The question is, do you want to be part of the legacy that restores our faith in humanity?

Although some of what I have to say will sometimes be hard to hear, I want you to know that all is not lost – there is still so much worth saving.

How bad we let things get is still up to us – the apocalypse is not a done deal.

PART 1

The Head

1
———

Elemental Earth

WHEN YOU WATCH THE SUNRISE from the beach, it's easy to be enchanted by the beauty of the elements around you. Light refracting over water, ribbons of clouds streaking through the sky like brush-strokes. It reminds us of the eternal ebb and flow of things, all ultimately driven by the sun. The world around us is alive, cycling through natural processes that have been operating for billions of years, eons before primitive humans evolved out of Africa some 300,000 years ago. To understand the climate crisis we are currently facing, it's important to first appreciate the building blocks of our planet. From there, we can easily see how human activity has disturbed the equilibrium that has maintained a safe operating space for humanity for thousands of years.

In its most elemental form, our planet is made up of land, life, water, ice and air. Together, these components form intricate cycles that are governed by the fundamental laws of physics, chemistry and biology. These five main systems, or spheres, all interact with one another to maintain the conditions we experience on Earth. The geosphere consists of the interior and surface of Earth, which are made up of hard rocks and softer sediments created by the processes of erosion driven by wind, rain and ice. The landscapes we see around us are the result of millions of years of rocks being uplifted, compressed and ground down, forming mountains, valleys, coastlines, canyons and everything in between. They are the backdrop of our world, our solid ground.

The limited part of the planet that is made up of living things is known as the biosphere. This includes ecosystems like forests, wetlands, coral reefs and deserts. Essentially, the biosphere is the Earth's life-support system. It provides us with the raw materials needed for human society to function: our food, shelter, transport, and industry are all sustained by an intricate web of life. Every living organism – from bacteria to bats – forms a unique component of broader ecosystems that function together to create our living planet. Aside from its utilitarian function, the natural world is also the source of our ancestry and kinship with all living beings. It is our primal sense of inspiration, joy and belonging.

But life on Earth would not be possible without the presence of water. The hydrosphere encompasses all the liquid water on the planet: oceans, rivers, lakes and groundwater. Seventy percent of the Earth is covered by water; the ocean accounts for 97 percent, while freshwater makes up less than 2 percent of all available water. When water molecules evaporate from the surface of the ocean or a leaf, they rise into the atmosphere, condensing to form clouds and eventually falling as rain or snow. The warmer the temperature, the more evaporation occurs, resulting in a wetter atmosphere saturated by more water vapor. In cold regions of the world, water freezes to form ice that blankets the poles and high mountain ranges. The frozen areas of our planet are collectively known as the cryosphere, taking in large ice sheets and glaciers that are crucial for maintaining the Earth's thermostat.

Above us, the atmosphere is an envelope of gases that keeps the planet warm, providing oxygen for life to exist and carbon dioxide to fuel photosynthesis in plants. Sometimes referred to as the great "aerial ocean," the atmosphere circulates around the globe in well-worn grooves that correspond with the location of landmasses, oceans and the presence of land surface features like mountain ranges and ice sheets. These tracks migrate north and south, and east and west as the seasons shift, altering the transfer of heat around the planet. The atmosphere itself is composed of about 78 percent nitrogen, 21 percent oxygen, and 0.9 percent argon. The remaining 0.1 percent contains trace gases including carbon dioxide, methane and ozone. Synthetic industrial chemicals like chlorofluorocarbons (CFCs), which were once used widely in refrigerants and aerosol propellants, are now also present, substantially modifying the chemistry of the atmosphere. These chemical pollutants have created a hole in our ozone layer – the part of the atmosphere that acts as a shield to protect the Earth from the sun's harmful ultraviolet radiation, which increases the risk of skin cancer, cataracts and immune system impairment.

The natural greenhouse effect is the warming of the Earth that results from the presence of heat-trapping gases like carbon dioxide, methane and water vapor in the atmosphere. Some greenhouse gases come from natural sources; for example, plants release carbon dioxide when they respire, methane is released from decomposing organic material, and water vapor evaporates from the land and water bodies. Water vapor is the most abundant greenhouse gas, but human activity contributes very little to its concentration in the atmosphere, so it is not a key driver of climate change. Without greenhouse gases, heat from the sun would be reflected back into space, cooling the surface of our planet to 18°C below freezing. It's the increasing concentration of greenhouse gas emissions from human activities like the burning of fossil fuels and the clearing of land that have altered the natural energy balance of our planet.

Since the Industrial Revolution began in England around 1750, human activities started to have a discernible impact on the Earth's climate and ecosystems. So much so that scientists have now named this period the Anthropocene – a geologic epoch dominated by humanity. This period saw the rapid increase of greenhouse gases and pollution particles (aerosols) in the atmosphere from the burning of fossil fuels like coal, oil and natural gas to support processes of electricity generation, transport and industrial activity. This dramatically altered concentration of heat-trapping gases is known as the enhanced greenhouse effect. Vast quantities of fossil fuels that were stored in the ground over eons of geologic time have been released back into the atmosphere at such rapid rates that they can't be absorbed fast enough by natural processes that lock up carbon in vegetation, soil, geological formations, and deep ocean trenches. Professor James Hansen, one of the world's most esteemed climate scientists, estimates that around a quarter of the carbon dioxide emitted from the burning of fossil fuels will "stay in the air for an eternity" – or more scientifically speaking, cycles through the Earth's complex systems for more than 500 years.

Widespread deforestation has also altered the land surface's ability to soak up excess carbon dioxide produced from humanity's relentless burning of fossil fuels. As more and more of the Earth's surface has been converted from natural vegetation ecosystems like forests or wetlands into agricultural crops or concreted urban areas, there are fewer living organisms like plants to extract carbon dioxide from the atmosphere during photosynthesis – the process of turning carbon dioxide and water into oxygen and glucose. This combination – the emission of pollutants through the burning of fossil fuels and the loss of vast tracts of forests – has resulted in the energy imbalance we are witnessing today. In a stable climate, the amount of energy that the Earth receives from the sun is roughly in balance with the amount of energy that is lost to space in the form of reflected sunlight and thermal radiation. Right now, human-driven increases in greenhouse gas emissions are interfering with this balance, causing the system to accumulate excess energy, which alters the Earth's climate. Unlike times in the past when only natural processes operated, our climate now contains an imprint of human activity that is imbalancing the Earth's natural systems.

At the same time, there has been an exponential increase in the size of the human population, from approximately 800 million in 1750 to 7.9 billion today, contributing to the colossal consumption of resources to support human activities. Since the 1950s, the Earth has entered the most pronounced period of the Anthropocene, now known as the "Great Acceleration." During this time, human activity has resulted in unprecedented rates of environmental change that have drastically altered our planet. For example, we now move more sediment and rock each year than natural processes like erosion and river flows. These changes are so profound that they will be permanently etched into the Earth's geologic records for millions of years to come. We have altered the Earth's natural cycles to the point where humans are now considered a geologic agent alongside erosion, volcanism and plate tectonics. In a single lifetime, humans have become a force of nature.

*   *   *

Before humanity exerted its dominance over nature, the Earth's climate was regulated by a range of natural processes operating on very long-term, geologic timescales. To understand how fast our world is changing, it's helpful to look back over the Earth's recent history. Over the past 2 million years – a period where we have very detailed geologic records – variations in the Earth's orbit strongly influenced the geographical distribution of sunlight, leading to the waxing and waning of great ice ages. Shifts in the orbit of the planet around the sun, variations in its tilt, and changes in the rotation of the Earth on its axis combine to influence surface climate conditions. Collectively, these orbital changes are referred to as "Milankovitch cycles," named after Serbian geophysicist Milutin Milanković, the first scientist to put forward these ideas as early as the 1920s. His theory recognizes that at our most fundamental level, we are a planet floating in space, influenced by other bodies in the solar system as we circle the sun. Small changes in our orbit influence the amount of solar energy reaching the Earth, where it lands and how intense it is. These cycles take place over thousands of years, ranging from 26,000 years to 100,000 years. So although they are not the primary driver of changes that human societies are experiencing right now, they form the deep time backdrop of our current climate.

These astronomical variations combine to determine the seasonal intensity of the sunlight reaching the poles. This is critically important for the growth and decay of ice sheets that influence the energy balance of the Earth. This is because snow reflects more solar radiation back into space compared with areas of open ocean which absorb more heat. Large continental ice sheets have expanded and contracted repeatedly in the past, drifting in and out of natural ice age cycles. Times in the Earth's history when the planet has been cool and covered in large ice sheets are known as "glacial" periods, and intervals of thawing and warming are called "interglacial" periods. During these cycles, average global surface temperatures varied by 5–7°C, with major changes in global ice volume, sea level and greenhouse gas concentrations. In the middle and high latitude of the Northern Hemisphere, temperature changes were as large as 10–15°C in some regions, highlighting the importance of the growth and decay of polar ice sheets in influencing the global climate.

Carbon dioxide concentrations in the atmosphere closely track temperature fluctuations, decreasing during glacial periods to around 180 parts per million (ppm), and increasing during interglacials to around 280 ppm. This is because the solubility of carbon dioxide is regulated by temperature, with colder water absorbing more carbon dioxide and warmer water less. In colder liquids, gas molecules move slowly, causing them to diffuse out of solution much more gradually, so more gas tends to remain in cold solutions. During glacial periods, the majority of carbon dioxide removed from the atmosphere, vegetation and ocean surface was mixed into the deep ocean and stored as calcium carbonate in marine sediments. Expanded sea ice also covered up regions of the upper ocean where upwelling returns carbon dioxide to the surface, reducing the amount of carbon dioxide released back into the atmosphere. When temperatures rise again during warm interglacials, the ocean starts releasing carbon dioxide back into the atmosphere, like a warm, carbonated drink going flat. The remainder of carbon dioxide present during these cycles is regulated by plants and soils, storing more carbon during cool periods and releasing it during warm periods.

The peak of the last ice age occurred around 20,000 years ago, long before modern human civilizations existed. Vast ice sheets covered 25 percent of the Earth's surface, cloaking many continents and ocean areas around North America, northern Europe and Asia (today, ice sheets now only make up around 10 percent of the Earth's surface). Forests and swamplands were covered in ice, trapping carbon and methane in frozen soils known as permafrost. These massive ice sheets also locked up vast quantities of water, lowering global sea level by around 120 meters. As the sea drained away, formerly submerged continental shelves fringing the coast became land bridges, connecting areas like Australia with Papua New Guinea through the Torres Strait, England to mainland Europe across the North Sea, and Siberia to Alaska via the Bering Strait. Humans, plants and animals mingled across geographical barriers as previously submerged land resurfaced in a reconfigured world.

Around 12,000 years ago, the melting of the ice sheets stabilized, marking the start of the current geologic epoch known as the Holocene. Since the recovery from the last ice age, global surface temperatures fluctuated little more than 1°C during this period, resulting in the stable climate conditions that allowed humans to thrive.

During warm interglacials – like the one we are in right now – ice caps retreat, sea levels rise, and forests begin to recolonize areas that were once blanketed by thick ice sheets. Vast areas of permafrost start to thaw, releasing carbon dioxide and methane from ancient forests and swamps that have been buried for thousands of years. We often hear more about carbon dioxide, but it's important to understand the role of increased methane emissions in accelerating climate change. Although methane only stays in the atmosphere for around a decade, it is far more potent than carbon dioxide – but more on that later.

For the first time in human history, these natural sources of greenhouse gases are being swamped by the cumulative load of anthropogenic emissions from the burning of fossil fuels, pushing the concentration of heat-trapping gases like carbon dioxide and methane into uncharted territory. As a result, atmospheric concentrations of methane are now higher than at any time in at least 800,000 years. Current carbon dioxide levels of 416 ppm are unprecedented in at least 2 million years, a staggering 1.7 million years before modern humans evolved. We are conducting a real-time experiment with our world – one that is rapidly taking us away from natural processes that shaped the entire course of the Earth's history, into an era dominated by humans.

*   *   *

I'm hoping this very distilled overview gives you a foundation for understanding how the Earth's history influences where we find ourselves today, and how human activities are now overwhelming natural processes. Geologic records over the recent past reveal that warm conditions are the exception rather than the rule. It's actually a really fascinating thing to consider – our time on this planet has been miraculously fleeting when considered in the context of geologic time. Over the past 430,000 years, the planet has only spent around 20 percent of the time in warm interglacial periods and around 80 percent in ice ages. The current warming began around 12,000 years ago, with peak warmth occurring around 6500 years ago, a period referred to as the Holocene thermal maximum. What is particularly interesting is that from that point onwards, the planet began to cool at a rate of roughly 0.15°C per 1000 years, ending abruptly when the Industrial Revolution began around 1750. The large-scale burning of fossil fuels resulted in a sharp spike in global temperatures, reversing this natural, long-term cooling trend, propelling us into an unprecedented era of widespread, rapid and intensifying human interference with the climate system.

When comparing today's conditions with natural climate variations like ice age cycles, it's important to understand that deep time changes happen very slowly – typically over tens of thousands of years. As humans, we are used to thinking in terms of days, weeks, months and years, not on geologic timescales that operate over thousands, millions or even billions of years. It's a lot to wrap your head around, but let me try to put things in perspective. It took the planet about 5000 years to warm around 5°C recovering from the height of the last ice age; that's a rate of 1°C warming every 1000 years, or 0.1°C per century. In comparison, global surface temperature has increased 1.2°C since industrialization, which is around seven times faster than the average rate of warming since the last ice age. Since 1970, warming has shot up further still, with global average temperature rising at a rate of around 1.7°C per century.

Similarly, carbon dioxide levels since the end of the last ice age have increased by around 80 ppm over 5000 years. Since then, concentrations have shot up from pre-industrial values of around 280 ppm in 1850, to 416 ppm in 2021 – a rise of 136 ppm in 171 years. That's fifty times faster than the increase following the last ice age. The IPCC presents multiple lines of evidence that indicate that the rate of increase in carbon dioxide in the atmosphere since 1900 is at least ten times faster than at any other time during the last 800,000 years – a period encompassing eleven warm interglacial periods. Atmospheric concentrations of methane, which are primarily released from the oil and natural gas industry, agriculture, waste, biomass burning and wetlands, are now 2.5 times higher than they were during pre-industrial times. Geologic processes are now playing out in fast forward, which doesn't give ecosystems and human societies much time to adapt.

When you step back and consider the long-term perspective of the Earth's natural climate variability, it's easy to see that the world is now warming very rapidly because of human interference. A comprehensive review published by the Geological Society of London in 2020 reported that the current speed of climate change is effectively without precedent in the entire geological record spanning 4.6 billion years. The only exception is the instantaneous meteorite strike that caused the extinction of the dinosaurs 66 million years ago. They conclude that while atmospheric carbon dioxide concentrations have varied dramatically in the geological record due to natural processes, the current rate of change is orders of magnitude greater than anything we have seen over the Earth's entire history.

So although the planet has experienced many natural climate cycles like ice ages in the past, it's the accelerated rates of change that are of most concern to scientists. Adapting to 1°C of global warming over 1000 years coming out of the last ice age is a vastly different prospect than adapting to an additional 2–3°C of warming over the lifetime of a human born today. It's a mind-boggling thing to consider. Our world is changing faster than the adaptation and resilience thresholds that many natural and human systems have evolved to cope with. When we compromise the integrity of the building blocks of our planet, we undermine the very foundation of human civilization and all life on Earth.

Ruthless pressure on the planet now finds us on the threshold of destabilizing the conditions that allow life to exist. In 2009, an international group of scientists proposed the concept of "planetary boundaries" to provide a framework for assessing humanity's global interference in the Earth's systems. These boundaries refer to values that are considered a safe distance from critical instability thresholds. Once breached, a system may become unstable and begin operating dangerously, with disastrous consequences for humans. Researchers have identified nine planetary boundary systems: climate change, land use modification, altered biogeochemical flows (phosphorous and nitrogen cycles), freshwater use, atmospheric aerosols, ozone depletion, ocean acidification, biodiversity loss, and the introduction of "novel entities" including toxic chemicals and plastics. So far, scientists estimate that four of the nine planetary boundaries have crossed into the danger zone: climate change, biodiversity loss, land use change, and human interference with biogeochemical cycles from intensive use of agricultural fertilizer.

New research led by Linn Persson from the Stockholm Environment Institute has proposed that the planetary boundary for novel entities has now also been breached. The study shows a fifty-fold increase in the production of chemicals for plastics, pesticides, consumer products and pharmaceuticals since 1950, with this number projected to triple by 2050. The production of plastic alone increased by 80 percent between 2000 and 2015. Around 60 percent of all plastics ever produced are still found in landfills or the natural environment, having very negative consequences for biodiversity through ingestion and entanglement and the contamination of land and marine ecosystems. Between 1950 and 2015, only 9 percent of all plastics were recycled worldwide. An estimated 350,000 manufactured chemicals are being produced and released at a pace that outstrips the capacity of governments to assess global and regional impacts on the environment. These horrifying statistics highlight not only how much relentless human consumption has degraded the natural world, but also the urgent need for the monitoring and regulation of the synthetic chemical industry worldwide.

Breaching critical planetary boundaries threatens the ability of the Earth system to maintain the climatic stability we've experienced for more than 12,000 years, a condition that allowed complex human societies and agriculture to develop. During this geologically stable period of the Holocene, environmental change occurred naturally and incrementally. There were also far fewer people using simple tools over smaller areas. But as the human population began to increase exponentially, so did the global use of fossil fuels and industrialized forms of agriculture, to the point where we are now destabilizing the systems that are essential for maintaining life on Earth. We have modified the Earth's surface and altered the chemistry of the atmosphere and oceans to the point where we have disrupted the equilibrium of a climate that has remained relatively stable for thousands of years. The results could be abrupt and, in some cases, irreversible.

We are the first generation of people to realize the gravity of bearing responsibility for maintaining a safe environment for humanity and the diverse life forms we share our planet with. The future of the world is in our hands.

2

The age of consequences

WHEN YOU ARE DEEPLY IMMERSED in daily life, it's hard to notice how incrementally things are changing, until one day, reality snaps sharply into view. Sometimes you notice it when you visit an elderly relative you haven't seen in a while and are stunned by how much they've aged. Or maybe it's the surprise appearance of gray hairs in the mirror despite how young you feel. Or perhaps it's the moment you realize that your child is no longer a baby, but a fully-fledged human in the world. Whatever it may be, it reflects the imperceptible passing of each day that eventually cumulates in a lifetime – a mix of the consequences of every choice we've ever made, and a solid dose of random chance.

Just like our lives, gradual changes in our climate are happening all around us, expressing the momentum of physical processes that have built up over the entire course of the planet's history. Aside from deep time fluctuations in the Earth's orbit, our weather and climate also vary on shorter, human timescales of days, months, seasons, years and decades. It's important to understand the difference between weather and climate, and how closely they are intertwined. Weather occurs from day to day on timescales of minutes, hours, days and weeks. The climate, on the other hand, is a long-term average of all daily weather experienced across a given month, season, year, decade or century. These background conditions accumulate into a dry month, a wet spring or a hot year. Typically, climate is considered over thirty-year periods, to provide a long enough baseline – or climatology – to determine the natural variability of a region. The main factors that influence natural climate variability on shorter timescales are variations in solar activity, episodic volcanic eruptions and large-scale ocean–atmosphere cycles like the El Niño phenomenon in the tropical Pacific and the North Atlantic Oscillation.

To understand the connection between the long-term warming of our planet and the extremes we experience from day to day or season to season, we first need to understand the processes that influence our weather and climate. The Earth rotates on its axis every twenty-four hours, giving us our night and day. The part of the globe directly facing the sun experiences daytime, while the shadow side is blanketed by the dark of night. The shape of the Earth and the intensity of sunlight at different latitudes create what is known as a thermal gradient – a distinct contrast in the temperature experienced between warm and cool areas – which sets up the conditions that circulate our atmosphere and ocean. Surface winds are generated by temperature differences: air rises over warmer land and ocean surfaces and drifts towards cooler areas, where it sinks in great convection loops. As oceans are warmed by the sun, water evaporates and condenses into huge cloud bands. As the humidity rises, excess moisture in the atmosphere eventually falls as rain, or as snow at high elevations or in polar locations. The presence of clouds, in turn, influences surface temperatures by trapping heat from the land surface.

Offshore, water is constantly flowing across huge ocean basins around continents and across the planet, upwelling cooler, saline waters from the depths to mix with the warmer, fresher surface waters where heat and carbon are released into the atmosphere. In turn, water masses cool as they flow towards the poles, sinking into the deep ocean where they can remain for decades or even centuries. This endless cycling of the ocean is known as thermohaline circulation – where *thermo* means temperature and *haline* refers to salt content. The process is driven by surface heat and the influx of freshwater from rain, river runoff and melting snow, which determine the density of seawater and how easily it sinks.

While some of the Earth's natural cycles have some regularity in their behavior, generally they are hard to predict because the exact physical conditions never repeat in entirely the same way – they are essentially chaotic, but their variations follow consistent themes. For example, we understand that factors such as latitude – the gradation of locations stretching from the equator to the poles – result in distinct climatic zones. At the equator, sunlight is received directly overhead over much of the day, giving these areas their consistently warm, humid climates. As you start to move towards the poles, solar radiation strikes the Earth's surface at more oblique angles and loses some of its strength, resulting in cooler climates. At such high latitudes, countries like Norway, Canada and Russia, which have areas that lie north of the Arctic Circle at approximately 66°N, can experience almost no darkness in summer, resulting in the phenomenon of the "midnight sun" when the sun remains visible in the dead of night. The opposite phenomenon, known as the "polar night," occurs in winter when the sun remains below the horizon throughout the day, plunging these areas into nights that last longer than twenty-four hours, with many shades of twilight. In contrast, at the equator, where the sun remains consistently overhead, the daytime length remains around twelve hours in all seasons. The difference between warm equatorial areas and the cool polar regions is what primarily drives the circulation of the atmosphere and ocean, redistributing heat around the planet from season to season. This is what forms the Earth's basic climatic zones.

Other factors like how far inland or how close to the coast a location is, and local features like the presence of mountain ranges, vegetation, water bodies or urban areas, influence a region's long-term climate. Elevated coastal areas generally receive the most rain, as they are located right near an endless supply of moisture from the ocean. Rain systems evaporate away as they travel over hot continental areas and rugged mountain ranges into arid zones where we find the great sandy deserts of the world in places like Africa, South America, China and Australia. In cities, the replacement of vegetation and soil with hard surfaces that absorb, store and radiate more heat results in hotter temperatures than surrounding rural areas, a feature known as the "urban heat island" effect. Together, these factors contribute to the local weather and climate conditions experienced at specific locations around the world.

Because the surface of the Earth differs so much from place to place, climate conditions in each region vary considerably, meaning that an increase in average global temperature will not be felt uniformly. For example, between 2011 and 2020 the average temperature over land increased by approximately 1.59°C above pre-industrial levels compared with 0.88°C over the oceans. Similarly, some regions will warm more rapidly than others, depending on their climatological features. The best-known example of this is "polar amplification," where warming is far more pronounced in polar areas of the Arctic compared to the rest of the globe. This is primarily because of the melting of ice and snow, which reveals darker land or ocean underneath, increasing the amount of solar radiation absorbed at the surface instead of being reflected back into space. As a result, annual Arctic temperatures have already increased at a rate three times higher than the global average, with a temperature rise of 3.1°C recorded between 1971 and 2019. Exceptionally strong warming has been observed during the cool-season months of October–May, when average temperatures across the region have increased 4.6°C, with peak warming of 10.6°C occurring over the northeastern Barents Sea. These are shocking figures to consider, as once we start losing large areas of ice from the poles, history tells us that the planet's climate can change drastically - and quickly.

*   *   *

Understanding how climate is changing all over the world is critical for understanding current and future risks to human societies and ecosystems. It's the reason why the IPCC conducts its global climate assessments - that way, we have an official, detailed inventory of how our climate is changing as the planet continues to warm, and clear guidance about what to do about it. The IPCC was established by the World Meteorological Organization (WMO) and the United Nations Environment Programme (UNEP) in 1988, to prepare comprehensive reviews and recommendations to all levels of government on the state of knowledge of the science of climate change; the social and economic impacts, and potential response strategies to minimize disruption to human and natural systems. Since the *First Assessment Report* was published in 1990, the IPCC has delivered six assessments, universally accepted as the global authority on climate change science. While very mindful of not being "policy prescriptive," IPCC reports provide the state-of-the-art scientific foundation that feeds directly into international climate policy making.

Some people argue that we don't go far enough in our assessments or are too conservative in our phrasing, but it's an unfair criticism that doesn't appreciate the role of scientists in complex decision-making processes. We are just scientists reporting our work as clearly as possible within the scope of our disciplinary standards - we don't wield the power of governments to enact policy. It's unreasonable to expect more of our community: the role of IPCC reports is to provide a rigorous assessment of the scientific consensus, not to offer political commentary. This doesn't stop authors sharing their individual perspectives outside of their IPCC role – like me writing this book – we just need to be clear that this is a personal view, and is not necessarily the consensus position of the IPCC report, which has undergone countless rounds of expert review to settle on the precise phrasing of each sentence.

As anyone directly involved in an IPCC assessment will tell you, the effort that goes into producing these reports is truly colossal. A typical volume of an assessment report consists of a dozen or so technical chapters coming in at around 80,000 words each. That's a report of well over 1 million words, not including technical appendices, glossary, lines of computer code to reproduce figures contained in the report, and a variety of other communication products developed to highlight key findings for the public. And there are three IPCC Working Groups, so that's done for all three volumes. Each chapter is then distilled down to a "Technical Summary" of around 150 pages for the *Sixth Assessment Report*, then condensed yet again into a forty-page "Summary for Policymakers." The latter is considered the most influential part of the report, so is made available in the six major UN languages – English, French, Chinese, Russian, Spanish and Arabic – and a range of other languages including German, Portuguese and Swedish.

The entire assessment is written by a team of international experts who volunteer their time – an act of altruism that needs to be seen to be believed. In recognition of this incredible effort, in 2007 the IPCC and former US vice-president Al Gore were jointly awarded the Nobel Peace Prize "for their efforts to build up and disseminate greater knowledge about man-made climate change, and to lay the foundations for the measures that are needed to counteract such change."

To quantify human and natural changes in the Earth's climate, a range of key variables are monitored using a variety of records collected by observing systems all over the planet. The World Meteorological Organization – a co-founder of the IPCC – uses a set of key indicators of temperature, greenhouse gas concentrations, ocean heat content, sea level, sea ice extent, glacier mass balance, and ocean acidification to monitor the state of the global climate. While changes in other variables like rainfall are critically important for human society, they vary considerably from region to region – one area might be drying while another is becoming wetter – so averaging these changes into a single global figure isn't always scientifically meaningful. As a result, alongside estimates that focus on global-scale changes, a range of regional indicators are also assessed in IPCC and other scientific reports to help individual countries understand the climate risk associated with their specific part of the world.

Things get complex quickly, because the physical processes that govern the Earth are so interconnected: a change in one parameter often has a knock-on effect in other aspects of the system. A good example is the pronounced warming in the Arctic, which has caused rapid and widespread changes in other indicators like sea ice and permafrost. For example, the extent of Arctic sea ice in September decreased by 43 percent between 1979 and 2019, with declines observed in overall coverage across all months. Sea ice is also becoming younger, thinner and faster moving as temperatures increase rapidly in the region. The latest IPCC report states that Arctic sea ice levels in summer are now the lowest they have been in at least 1000 years. All these indicators are a clear sign that the Arctic is undergoing fundamental change. Because of the importance of polar areas in regulating the Earth's climate, we know that these rapid changes have far-reaching consequences for the rest of the planet.

As ice also recedes from land areas of the Arctic, underlying permafrost has warmed by 2–3°C since the 1970s. This is releasing previously locked-up carbon dioxide and methane into the atmosphere, reinforcing warming feedbacks that affect atmospheric greenhouse gas concentrations. Permafrost underlies around a quarter of the Northern Hemisphere land surface and stores an estimated 1700 billion tonnes of carbon in frozen ground, which could be released into the atmosphere as the planet continues to warm. Arctic permafrost contains a massive frozen store of ancient organic carbon, equivalent to approximately twice the amount of carbon that is currently stored in Earth's atmosphere. This carbon has accumulated over tens of thousands of years during past ice ages, when dead plants were buried and trapped within layers of frozen soil. Prevailing cold conditions prevent ancient organic material from decomposing until conditions warm up and the landscape begins to thaw. As temperatures increase, organic matter in these soils begins to decompose and return to the atmosphere as either carbon dioxide or methane, which are both important heat-trapping gases. Methane has around eighty-six times more warming power than carbon dioxide averaged over twenty years and is thirty times more powerful when averaged over 100 years, so any major releases from these areas will significantly amplify future warming.

As global warming continues, we face the threat of converting Arctic permafrost into a net source of carbon, rather than a carbon sink. The latest IPCC assessment states that it is virtually certain that the extent and volume of permafrost will shrink as the climate warms. It estimates that the volume of permanently frozen soil within the top 3 meters of the ground will decline by 25 percent for every 1°C of global warming. As previously frozen land starts to thaw in places like Canada, Russia and Alaska, extensive areas of the land surface have already begun to subside, buckling roads, cracking buildings and threatening the viability of many airports. These areas of collapse can expose deeply buried permafrost, further accelerating thawing. The situation is being made even worse by the increasing frequency and severity of wildfires in the Arctic, emitting enormous amounts of carbon both directly from combustion and indirectly by accelerating permafrost thaw. As the climate warms, Arctic wildfires are projected to increase 130–350 percent by mid-century, releasing above-ground biomass and an increasing quantity of ancient permafrost carbon that has been dormant for thousands of years. In 2020, over 50 percent of Arctic wildfires burned in ice-rich permafrost areas considered to contain the most carbon-rich soils in the Arctic – an unusual occurrence that scientists fear will accelerate the rate of thawing and release of carbon in these areas.

Although these processes amplify warming, they are not included in the vast majority of climate models, so we don't know precisely how they will influence future warming. That said, recent climate modeling experiments run with the rapid release of methane and carbon dioxide from the Arctic are starting to give us some clues. Releasing just a fraction of these sources into the atmosphere will make climate change happen faster than is currently accounted for by most models. The concern is that permafrost carbon emissions will be felt over coming decades and centuries, taking us beyond what models indicate we can expect from the burning of fossil fuels alone.

Despite growing evidence of increased carbon loss from permafrost regions, the IPCC states that it is currently not possible to draw any global quantitative conclusions based on ecosystem- or site-specific results, as each ecosystem has complex seasonal behavior that is still very difficult to model. Based on current observations, the IPCC concludes that future permafrost thaw will lead to some additional warming – enough to be important, but not enough to lead to "runaway" warming, where permafrost thaw would lead to a dramatic, self-reinforcing acceleration of global warming. Right now, it is only possible to say that the warming is strong enough to be included in estimates of remaining carbon budgets that determine the level of warming the world will experience, but weaker than the warming from the burning of fossil fuels, which is by far the dominant factor driving global warming.

That said, it's hard to really know what the future holds based on the wide range of published estimates and our incomplete understanding of the timing, magnitude and predictability of permafrost feedbacks. It's also unclear whether the permafrost carbon pool represents a widespread global tipping point with a single abrupt threshold at a given level of global warming, or multiple regional tipping points. At this stage, it's too hard to tell how much of the change will play out gradually, and how much might occur in sudden, unpredictable ways. But what we do know is that abrupt change has happened in the past on regional scales and cannot be ruled out based on our current understanding. The science clearly shows that the risk of initiating these feedback loops increases with higher levels of warming, so it's something we should do everything in our power to avoid. We know that once these invisible thresholds are breached, they are considered irreversible for decades and centuries to come. The problem is, we will only find out when enough definitive studies show that the signal has clearly emerged above the background of natural climate variability. By then, the genie will be well and truly out of the bottle.

*   *   *

We know that the planet is warming, but just how quickly and dramatically things are changing is stunning the scientific community. In June 2021, the Pacific northwest region of the United States and southwest Canada experienced extreme temperatures that shattered historical temperature records by such extraordinary margins that some experts believe the event could be the most extreme heatwave in modern history. Many cities in the western United States regions of Oregon and Washington and the western provinces of Canada recorded temperatures well above 40°C (104°F), including a phenomenal new national temperature record of 49.6°C (121.3°F) on 29 June in Lytton, British Columbia – a village located 150 kilometers northeast of Vancouver. It broke the all-time Canadian heat record of 45°C, set in Saskatchewan in 1937, by an incredible 4.6°C. Usually, temperature records are exceeded by tenths of a degree, not by such enormous margins. This is not just breaking a record;

it is an obliteration. Never in the history of weather observations have so many all-time heat records fallen by such large margins.

It's the first time we have observed such high levels of warming so far poleward anywhere on the planet. We are talking about desert-like heat in Canada – heat that would be considered extreme in some of the hottest places on the planet like the Middle East, Central Australia or southwest USA. To record close to 50°C at a sub-Arctic latitude of 50°N, close to the town of Whistler, one of the largest ski resorts in North America, is truly terrifying. It was so shocking that many experts were left speechless, with meteorologist Claire Martin from Environment Canada – the country's national weather service – simply tweeting "Words fail me."

For context, the record set in Lytton is hotter than any temperature ever recorded in Alice Springs in Central Australia (45.7°C) or the notoriously hot US desert city of Las Vegas (47.2°C). It's hotter than any temperature observed in Europe or South America, and the most extreme heat observed north of 45°N latitude. The highest temperature ever reliably recorded on Earth was 56.7°C (134.1°F) on 10 July 1913 at Death Valley in California, located at 36°N in the Mojave Desert – one of the hottest places on the planet, along with the scorching deserts of the Middle East and Africa. So for a place like Canada to be experiencing such extraordinary heat so close to the North Pole challenges our contemporary understanding of heatwaves. A tweet by Scottish meteorologist Scott Duncan summed up how many of us felt: "I didn't think it was possible, not in my lifetime anyway . . . This moment will be talked about for centuries." But the truth is, it's only a matter of time before these extreme temperature records are reset. Extreme conditions will start to become average sooner than we think.

Hot on the heels of extreme heat come extreme fire weather conditions. Just a handful of days after setting Canada's national temperature record, Lytton was destroyed by a ferocious wildfire, with close to

90 percent of buildings burnt to the ground. According to the mayor, the fire spread within fifteen minutes of the arrival of smoke, leaving the terrified community unprepared. The exceptional heat led to over 500 sudden deaths in the region and to sharp increases in hospital visits for heat-related illnesses and emergency calls in areas that are unaccustomed to such extreme temperatures. As many homes in British Columbia do not have air conditioners, many people fled to hotels or emergency "cooling centers" to find relief from the heat. According to local experts, the only event that compares is the July 1936 heatwave that occurred during the multi-year Dust Bowl drought in the Midwest of the United States and south-central Canada. In the days before widespread air conditioning, the impacts of the protracted heatwave were huge: there were over 5000 heat-related deaths reported in the United States and a further 780 in Canada after an exceptionally hot and dry summer during an extended drought.

For many of us in the climate field, the ferocious 2021 heatwave was yet another disturbing sign of how extreme and unpredictable our climate has become. If we can break records by such huge margins with just over 1°C of global warming, what will the world be like when we warm by 2, 3 or even 4°C? It makes me wonder if one day, climate scientists will look back at the 2020s and recognize it as the beginning of the end – the moment humanity finally tipped the balance and destabilized the Earth's climate.

*   *   *

In the aftermath of unprecedented extremes, people want answers. So exactly how do scientists determine how much human-caused climate change has influenced an individual heatwave, wildfire, flood or severe storm? How much is caused by natural fluctuations, and what proportion is being influenced by human modifications to the climate?

Scientists can quantify how the probability of an individual extreme weather event has changed due to human influences or natural factors by comparing results from global climate models with long-term observational records. The most common approach is to calculate the likelihood of events occurring in model simulations run with and without greenhouse gases from human activities. This area of climate science is known as "detection and attribution"; it's a relatively new area that really took off during the mid-1990s. It's a little like the field of forensics – but instead of analyzing human fingerprints, scientists examine the statistical fingerprints of well-described climate influences that match our database of known suspects.

The term "detection" refers to the emergence of a signal from background noise, demonstrating that the climate has exceeded what is possible from natural variability alone, in some defined statistical sense. "Attribution" is the process of determining the most likely physical causes for a detected change. State-of-the-art climate models that incorporate natural factors like solar, volcanic and ocean–atmosphere variability are compared with simulations that also include anthropogenic factors including greenhouse gas, ozone and land use changes. The two model scenarios are run separately – simulating a world with and without human influences – to quantify the contribution of natural variability and human activities on extreme events observed in weather and climate records.

Since the 1990s, a rapidly increasing number of studies have shown that anthropogenic greenhouse gas increases have influenced various aspects of global and regional climate change. Using instrumental climate observations from all over the world, researchers have shown that global temperature observations can only be reproduced by models that include human influences. When compared with models run using only natural factors, the two curves diverge. When greenhouse gases are added in, modeled and observed temperatures line up neatly. Scientists can now demonstrate that direct temperature observations match the fundamental thermodynamics simulated by a large suite of independent climate models. The implications are powerful: they definitively show that the climate is no longer being influenced by natural variability alone – humans are now directly altering the behavior of our climate.

Evidence of anthropogenic influence on the climate system has strengthened over the course of the IPCC's six assessment reports. When the *First Assessment Report* was released in 1990, there was little observational evidence for a detectable human influence – the signal had not emerged clearly from the background "noise" of natural variability based on the limited studies available at the time. However, that first report accurately predicted that "the unequivocal detection of the enhanced greenhouse effect from observations is not likely for a decade or more." When the *Second Assessment Report* was published in 1995 there was enough evidence to conclude that "the balance of evidence suggests a discernible human influence on global climate." But it was really the *Third Assessment Report* of 2001 – some twenty years ago – that found that a distinct greenhouse gas signal was statistically detectable in the observed temperature record and that "most of the observed warming over the last fifty years is likely [66–100 percent probability] to have been due to the increase in greenhouse gas concentrations."

In 2013, the *Fifth Assessment Report* concluded that "warming of the climate system is unequivocal" and that it was "extremely likely" (95–100 percent probability) that human activities are "the dominant cause of the observed warming since the 1950s," and "virtually certain" (99–100 percent probability) that natural variability alone cannot account for the observed global warming. The *Sixth Assessment Report* has amassed even more evidence to further support "unequivocal" human influence on the atmosphere, ocean and land since pre-industrial times, emphatically concluding that "human influence on the climate system is now an established fact." The proof is now so indisputable that it is considered a factual statement based on overwhelming evidence that no longer requires any of the IPCC's usual likelihood statements, including probabilistic qualifiers. It's like stating that the sky is blue or the Earth is round. This historic development is the result of insights gained from the collation of longer observational datasets, improved geological evidence, advances in climate modeling and statistical techniques and, perhaps most alarmingly, a stronger warming signal that has emerged since the last IPCC report was released in 2013.

Before the development of the field of attribution science, it was difficult to definitively link a specific weather event to climate change. In fact, it used to be a favorite talking point of climate change deniers, who failed to grasp the connection between changes in daily weather extremes and a warming planet. There is a lot of natural climate variability in the climate system, particularly when considering rainfall, which can vary considerably from location to location. That's why it's important to be able to quantify whether an event lies within or outside the range of past events seen in historical records. Another issue is that long records don't exist for some parts of the world, making it difficult to estimate the full range of natural variability. But major advances in climate modeling, computing power and the recovery of historical climate observations now allow multiple simulations to be run far more quickly and efficiently than ever before. We can then compare these simulations with a range of newly consolidated long-term climate records, improving our understanding of past weather and climate variability.

Now that scientists can statistically compare changes in the likelihood of extreme events using model simulations that are run with and without human factors, we can quantify how unusual or not a specific weather event may be. The truth is that all weather events are now occurring in a climate that is significantly hotter than it was fifty years ago, which means that *all* weather is now influenced by human-caused climate change. So unlike twenty years ago, the question is no longer *if* climate change is influencing extreme events, but by exactly how much? The answer is so extensive that, for the first time, the IPCC has dedicated an entire chapter to the attribution of extreme weather and climate events.

To help people make the connection between extreme weather and a warming climate, the World Weather Attribution team, an international group of leading extreme event attribution experts co-ordinated by researchers from the University of Oxford, was established in 2014 to provide quick, scientifically reliable information on how weather extremes may be affected by climate change. Part of their motivation is to use published, peer-reviewed statistical methods to provide a rapid assessment of the role of human influence on real-time weather extremes before public interest and the media coverage have moved on. Scientific papers typically take at least a year to publish, and by then it is often too late to contribute to public discussions when the impacts of climate change are clear for all to see. In an era when the stakes are so high, scientists are doing their best to make the connection between destructive weather events and climate change as clear as possible as quickly as they can.

In the aftermath of the exceptional heatwave in the Pacific northwest region of the United States and southwest Canada in June 2021, the World Weather Attribution team sprang into action, working around the clock to analyze the extraordinary conditions. Based on a thorough investigation of observations and climate model simulations, they concluded that the heatwave was virtually impossible to reproduce without human-caused climate change. They calculated that human influences made the heatwave at least 150 times more likely than possible by natural factors alone. The temperatures recorded during the heatwave were so extreme that they fell well beyond the range of historically observed extreme temperatures. They were so far outside anything ever recorded for the region that it was difficult to accurately estimate how rare the event actually was. The researchers conservatively concluded that it was likely to be a one-in-1000-year event based on observations of today's climate.

The team emphasized that in a future world with 2°C of global warming, which the IPCC estimates could be reached as early as the 2040s, such an extreme heatwave could happen once every five to ten years and be another degree hotter. Perhaps most disturbingly, the researchers warn that the climate system may have crossed a threshold that increases the probability of such extreme heat occurring outside of the gradual increases in warming that have been observed so far – that is, there are non-linear processes that are currently not being captured by the climate models. If this is true, we may be seriously underestimating how deadly global warming will be in the future. These results are yet another clear warning that as the planet continues to warm, exceptional heat will occur far more often and play out much sooner than we think. The authors conclude by saying:

> Our results provide a strong warning: our rapidly warming climate is bringing us into uncharted territory that has significant consequences for health, wellbeing, and livelihoods. Adaptation and mitigation are urgently needed to prepare societies for a very different future ... measures need to be much more ambitious and take account of the rising risk of heatwaves around the world, including surprises such as this unexpected extreme.

We can't say we weren't warned.

*   *   *

The reality of the climate crisis is hard to grasp. In part, this is due to the difficult process of sifting through mountains of technical information, especially IPCC reports, which are the grueling pinnacle of our exacting science. We assess, revise, reconsider and reassess once more. Every single word is agonized over by hundreds of experts trying to convey the nuance of our science with absolute precision and transparency. In the end, our words often appear impenetrable, weighed down by a burden of proof that does not plague other disciplines. A study published by social scientist Professor Elisabeth Lloyd from Indiana University in 2021 concluded that, compared with the level of evidence required by legal, regulatory, or public policy processes to establish proof "beyond a reasonable doubt," climate science demands too much of itself, setting the evidence bar way too high, given the imminent level of threat to society.

For example, the IPCC requires a probability of 90–100 percent before assigning a scientific claim, like the attribution of a specific weather event to human-caused climate change, as "very likely." In contrast, the legal profession requires demonstration of proof at a much lower level. For example, in the United States, the standard for a civil case in medical malpractice or patent infringement is "more likely than not," generally interpreted as a probability of more than 50 percent. Lloyd's study recommends that the IPCC should meet the "more likely than not" standard of proof required in court rooms, where incomplete information doesn't stop complex decisions from being made.

The same approach of responding cautiously in the absence of complete information was also applied by the medical profession during the COVID-19 pandemic. Incredibly fraught decisions had to be made, like what level of evidence is required to start actively preparing for catastrophic stress on intensive care units in hospitals? Or at what point do you decide to shut down entire cities to stop the spread of a deadly disease? When there is an immediate and grave threat to society, the precautionary principle asks us to proceed wisely, acting cautiously with imperfect evidence to protect the common good.

In the case of climate science, it's also worth remembering that there is a long history of vested interests in the fossil fuel industry that have sought to undermine public confidence in our work. As the largest exporter of coal in the world, the political debate in Australia has been particularly toxic and adversarial. The same holds true in the United States, also in the world's top five coal exporters, where political divisions around climate policy run deep. Consequently, conservative governments all over the world have demanded a disproportionate level of evidence to prove that human activities have significantly altered our climate before being willing to politically address the issue. This has resulted in a ruinous delay in our global response, landing us in the emergency we find ourselves in today.

The most confronting thing about the IPCC's *Sixth Assessment Report* is that the situation is now so bad that you don't have to dig too deeply to get a sense of the true scale and magnitude of the problem. Many of the findings have simply strengthened as the body of evidence has grown exponentially in recent years. In a nutshell, our report shows that recent changes in the climate are widespread, rapid, intensifying, and unprecedented in thousands of years. We explain that climate change is already affecting every continent and ocean on Earth, and these changes will escalate with further warming. Perhaps our most significant conclusion is that it is now an *established fact* that human activities have altered all aspects of the climate system. Of the 1.09°C of surface warming that the world has recorded since pre-industrial times (estimated from conditions observed from 1850–1900) until 2020, 1.07°C is due to all human activities, including the burning of fossil fuels, deforestation and the presence of aerosols in the atmosphere. In other words, humans are responsible for *virtually all* observed global warming. It is the cumulative impact of every decision ever made to exploit the natural world, the legacy of rampant exploitation of the Earth, with no end in sight.

Given that greenhouse gas emissions are still rising, understanding how the planet has responded during other warm periods in the Earth's history has become a global research priority. The study of past climates provides us with a long-term perspective on current conditions and what might lie ahead. Geologic records reveal that our world is changing at rates unseen for thousands – and in some cases millions – of years. For example, 2011–2020 was the warmest decade recorded in the past 125,000 years, the peak of the Earth's last global interglacial period. During this time, average global temperatures were 0.5–1.5°C warmer than pre-industrial conditions, and global sea level was 5–10 meters higher than it is today. The West Antarctic ice sheet collapsed, adding 3 or more meters of sea level rise to contributions made from the disintegration of the Greenland ice sheet alone.

The last time carbon dioxide levels were similar to present conditions was around 2 million years ago, a period in the Earth's history known as the Pliocene. During this warm period, global temperatures were 3–5°C warmer than pre-industrial conditions, which is three to five times the warming we have already experienced so far. During this period, the oceans around the Arctic Circle were 7°C warmer than pre-industrial conditions. Arctic tundra regions, which currently blanket vast permafrost areas, were warm enough to support forests that extended up to 2500 kilometers further north over Canada and 250 kilometers into Siberia, where summer temperatures were up to 6°C warmer than present. The Greenland and Antarctic ice sheets retreated substantially, unleashing 5–25 meters of global sea level rise that lasted thousands of years.

In contrast to these ancient warm periods, so far the current warming has resulted in a 20-centimeter rise in global sea level since the start of the twentieth century. Until recently, most of this was associated with the thermal expansion of seawater as it warms. But the latest IPCC assessment shows that the retreat of glaciers and melting of the Greenland and Antarctic ice sheets has eclipsed this process since the 1960s, now accounting for around 70 percent of the increase recorded between 1901 and 2018. The problem is, the majority of modern glaciers are still losing mass as they adjust to current warming, so are now wasting away at an alarming pace. For example, the loss of ice recorded from the Arctic and Antarctic ice sheets during 2010–2019 was four times greater than during the 1990s, indicating a rapid acceleration in recent years. The loss of so much ice from our planet means that we are already committed to a cascade of changes – even if we manage to stabilize our greenhouse gas emissions – as the world's oceans reconfigure to increased influxes of meltwater, altering the behavior of ocean currents that distribute heat around the planet. This process is now irreversible and will go on for centuries.

The IPCC's *Sixth Assessment Report* states that it is virtually certain (99–100 percent probability) that global sea level will continue to increase over the twenty-first century. As it stands, sea level is expected to rise an additional 10–25 centimeters by 2050 whether or not greenhouse gas emissions are reduced. Beyond 2050, further increases in sea level depend on the level of greenhouse gases we emit. In the worst-case scenario, global sea level will likely increase by 0.63–1.02 meters relative to the 1995–2014 average by the end of the century. But given how difficult it is to monitor and model complex ice sheet destabilization processes in remote polar regions, the IPCC suggests factoring in an additional 1-meter rise on top of the likely projected range by the end of this century.

How much sea level increases in a given location around the world depends on a number of factors including the shape of the sea floor, land subsidence in densely populated river deltas, local groundwater extraction, and the presence or absence of coastal protection infrastructure like sea walls and levees. These influences can result in regional sea level changes that can be as much as 30 percent lower or higher than changes in global mean sea level. Some areas most vulnerable to ongoing sea level rise and coastal flooding include the sinking cities of Bangkok, Jakarta, Tokyo, Shanghai and New Orleans. Many of these areas are classified as megacities – defined by the United Nations as a city with a population of more than 10 million people – representing a major threat to some of the most densely populated regions of the world. Where these people will retreat to needs to be strategically planned to avoid the mass displacement of people that may lead to social unrest as climate refugees seek higher ground.

Even under an intermediate-emissions scenario, Australia's sandy shorelines are projected to retreat by around 110 meters in eastern Australia and up to 90 meters on the south coast by the end of the century. Under a worst-case scenario, we see 100 meters of our coast disappear by 2100 all around the country, with retreats as high as 220 meters in northern Australia and 170 meters in eastern Australia, where most of the Australian population lives. Aside from the loss of large chunks of our coastline, severe storm surges are expected to further exacerbate coastal erosion, meaning we will witness many of our beloved beaches degrade and eventually disappear during our lifetimes. As someone who lives in a coastal town and visits the beach most days, it is heartbreaking to realize just how much the Australian way of life is already eroding before our eyes. I often think about the loss our Indigenous communities must feel as they witness their sacred lands struggling to survive the onslaught inflicted upon the Earth by a single generation.

The confronting truth we must face here is that past greenhouse gas emissions have already led to unavoidable future changes in the Earth's climate. One of the key conclusions of the IPCC report is that there is no going back from some changes, but others could be slowed and even stopped by limiting warming. How bad we let things get is still in our hands, but fast slipping out of reach. Even if the world's greenhouse gas emissions miraculously dropped to zero tomorrow, it will take time for the planet to regain its equilibrium in response to past emissions. It's what scientists refer to as "committed" warming: irreversible changes in the ocean, ice sheets and sea level that will play out until at least 2300, even under low greenhouse gas emissions scenarios. The IPCC estimates that by 2300, global mean sea level could rise by 0.3–3.1 meters under a low-emissions scenario. Under a very high emissions scenario, the likely increase is estimated between 1.7 and 6.8 meters by 2300, with an upper end as high as 16 meters if widespread instabilities lead to major ice losses around Antarctica.

Such extreme increases in sea level will completely reconfigure maps and life on our planet, in ways that people can't yet imagine. A 2019 study by risk analysts from the US organization Climate Central estimated that 1 billion people currently occupy land less than 10 meters above present high tide levels, including 230 million people below 1 meter. The risk of trying to rehouse coastal populations on such a monumental scale, while trying to protect critical infrastructure including roads and airports alongside fragile ecosystems, must be weighed up right now, while we still have time to refuse locking in this apocalyptic future.

*   *   *

If you've made it this far, I'm sure you'll agree that thinking about climate change can be overwhelming. Thinking about it day in day out as a scientist can sometimes feel so crushing. I often use my quiet moments to write in my journal, to figure out how I'm feeling. It helps me process the stress of a job that sometimes feels very alienating. Acknowledging that the world as we know it is coming apart is an act of courage. Once faced, it taints everything. Sometimes it's hard to put aside what I carry and live a carefree life. Every decision feels fraught; it's a constant dilemma between choosing to be a dedicated scientist trying to protect the future of our planet over living a simple, unencumbered life right now, while I still have age and good health on my side. It's confronting to realize how often I lose touch with myself and my friends, as the endless demands of my working life are met at the expense of my inner world and time with the people I love. These days, writing is some of the only time I get to slow down for long enough to hear my own thoughts. I can stop – not rush – and just be still and listen. Because my sleep is often so broken, my reflective time is precious; like many writers, it's how I restore myself, how I integrate and make sense of things.

Because of the relentless demands on my time, I'm often stretched to breaking point. Sometimes it's hard not to feel frustrated with the people around me – complaining students who don't want to leave their bedroom (can you put everything online so I don't have to come to class?); family who can't figure out why a woman my age doesn't have children (maybe you could just have one?); or friends who don't understand why I start crying while out hiking (it's just a fire, rainforests grow back, right?).

People don't really understand that I'm often preoccupied by other things, internally grappling with questions like: will the Great Barrier Reef be dead within ten years? Has the West Antarctic ice sheet become so destabilized that extreme sea level rise scenarios are the most realistic outcomes to consider? How bad are we going to let things get? I admit it's pretty heavy stuff, but there's no escape when your work forces you to think about these things. I don't have the luxury of turning away; looking into the void is my day job. I face these things not because I find it easy or comforting. I face them because I must.

Increasingly I'm finding that facing this reality with other people really helps. In November 2019, as the most punishing drought in Australian history neared its horrendous peak, I found myself completely burnt out after a long year juggling relentless commitments: starting a new job in a different city, setting up a research team, addressing hundreds of IPCC chapter revisions, developing and teaching two new university climatology courses, responding to crisis-mode media, all while doing a steady stream of public speaking events. A highlight was the Varuna Writers' House Mick Dark "Talk for the Future," an important annual address on the environment, delivered to a full house in Wentworth Falls in the Blue Mountains west of Sydney. I had driven up through terrifying brown plumes of bushfire smoke that saw catastrophic bushfire conditions declared for the first time in Sydney's history. Many of the people in the room that night had been evacuated from their homes just days before. It was touch and go whether the event would even go ahead; but there they were, turning out in droves even though the conditions outside were so volatile.

The atmosphere that night was electric. I'd never spoken to such an engaged audience at such an emotionally charged, historic moment; they understood the truth of my message from their own lived experience. As I spoke, we collectively joined the dots between the future projections and the painful reality of the crisis unfolding around us. Ancient ecosystems gutted, countless animals destroyed, inconceivable political denial of the science. It was the first time in my career I teared up at the end of a talk – there I was, speaking directly into an unprecedented national crisis; there was nothing more I could do as a scientist or as a human being to help people understand the scale of the emergency unfolding around us.

That night, my professional armor was dissolved by the power of humans coming together in a room to bear witness to each other's pain and grief. Sensing my emotion, the audience rose to their feet, giving me a resounding standing ovation I'll never forget. It was one of the greatest gifts of my career: acknowledgment that I am not alone. That there are many switched-on, kind-hearted people who really do care about the destruction of our Earth. Knowing that some people really are listening makes all the personal sacrifices I make as a climate scientist feel worthwhile. It was a reminder that there is still so much good in humanity, often thriving at the grassroots where all social change begins. We will come back to this later, but for now I just want to say that sometimes it's easy to focus on the people making things worse, overlooking all of the incredible people doing everything they can to make the world a better place. Our nurses on the frontline of the COVID-19 crisis, volunteer firefighters protecting our precious places, emergency workers facing untold horrors, our teachers doing all they can to give the next generation a fighting chance. My IPCC colleagues working thousands of unpaid hours throughout a deadly pandemic to finish our report. You choosing to listen by reading this book.

These are my people.

When I finally got a chance to retreat back into the oasis of my journal after the event, I tried to process the future I'd just glimpsed:

> By the time the next [Seventh] IPCC report comes out [around 2030], I fear it will be too late to have stopped the domino effect from kicking in. If we don't put the brakes on now, then really there is not much hope left. Instead, we will see a gradual worsening of conditions on the planet, more extreme events, more people displaced, and safe havens like Australia will soon be inundated by people just trying to stay alive. We've dealt with political instability in the past, but it's hard to imagine that climate change will be the next great force to shape the course of human history. Right now, battles are being fought on all fronts – politically, in our communities, by our business leaders, our scientists, all over the world. During the time of this

great turning, it's easy to lose hope. With so much loss, destruction and suffering, it's sometimes hard to imagine that we could pull back from this in time to avert a horrific cascade of destabilisation that will reconfigure the planet and human life as we know it. It's an extraordinary time to be a climate scientist on the frontline as the end of the world unfolds.

3
___

Fork in the road

SOMETIMES PEOPLE ASK ME HOW I came to be a scientist: was it a clear path I always knew I'd take, or was there a specific fork in the road? Unlike some of my colleagues, it's not something I knew I wanted to be early on. While I confess to owning a chemistry set, as a child reading was my favorite activity – I devoured everything I could get my hands on, disappearing for days on end into the world of books. As the daughter of Egyptian migrants, English, Arabic and French were spoken at home, so I readily took to languages at school and was fascinated by the different ways cultures express themselves. By around age seven, my teachers recognized my knack for creative writing, encouraging me with class awards and extension activities. There I learned that people who wrote books were called "authors," instantly resolving the answer to the question of what I wanted to be when I grew up.

But by the time I got to high school, I started to develop a real fascination with science. I had a string of great teachers who opened my eyes to the intricacies of the natural world and soon, my favorite subject was geography. I became captivated by how the Earth worked; the way landscapes formed, how to read a synoptic chart and why rainforests create their own rainfall. But for me, understanding physical processes was only half of the story. We also learned how human culture interacted with the natural environment: the impact of megacities on river floodplains, the bulldozing of biodiversity to make way for cash crops, the custodianship Indigenous people felt for their land. We were challenged to think about the balance between economic development and maintaining thriving ecosystems and diverse cultures. To me it felt like the most important thing to try and figure out: how humans can collectively overcome poverty and inequality, and live meaningful lives without trashing the planet.

In January 1994, a year before I finished high school, Sydney was engulfed by ferocious wildfires that ripped through bushland not far from my home. It was a terrifying experience: plumes of acrid smoke filled the sky, ash fell over my neighborhood like confetti. The fires of that summer caused the mass evacuation of thousands of people, claimed four lives, destroyed 225 homes and burnt through 800,000 hectares. Over 20,000 volunteer and professional firefighters fought the blazes, making it the largest firefighting effort in Australian history at the time. The disaster led to the formation of the New South Wales Rural Fire Service – the world's largest volunteer fire service – to unify the emergency response of the network of local fire brigades. In the aftermath, I wanted to understand more about the impact of the fires, so I developed a student project to investigate the extensive damage done to the Lane Cove National Park, a patch of urban bushland around 15 kilometers northeast of central Sydney. I spoke to people about how they felt about the fires and the destruction of the well-loved park. They responded with shock about damage to their local area, how close they had come to losing their homes and their sadness for the blackened forest, now eerily quiet.

For me as a sixteen-year-old, it was the first time in my young life that climate change injected itself into my reality; danger was now literally on my doorstep. It made me want to understand what drives Australia's fierce climate and what it meant for my own future. This curiosity led me into an environmental science degree, where I specialized in physical geography and learned foundational climatology. Things clicked; I took every climate subject on offer, as I began to understand how climate variability was the ultimate driver of all of life on Earth. In 1997, during my second year of university, one of the most powerful El Niño events in recorded history unfolded, unleashing a global wave of destruction in its wake. The theory I was learning in lectures dramatically sprang to life – an unusual warming of the equatorial Pacific Ocean led to a cascade of natural disasters that engulfed the world, from massive flooding in Latin America and the Horn of Africa, to severe drought in Southeast Asia and the southwest Pacific. Crops failed, forests burnt, and waterborne diseases proliferated as the "El Niño of the Century" swept the globe. It was a real-time example of how interconnected the climate system really was. Seeing images of widespread destruction once again reminded me how quickly the world changes with the weather – not just at home in Australia, but right across the planet.

During the final year of my university degree, my favorite lecturer encouraged me to apply for a PhD scholarship. I really wanted to learn more about El Niño and how unusual these events were in a long-term context – but by then, I was ready to take a break. I'd been studying for a long time and had recently fallen deeply in love, so my mind was now preoccupied by other things: we wanted to get to know each other better by seeing the world together. I'd traveled solo around Australia, Indonesia, Malaysia and Brunei in between semesters while doing my degree. I'd snorkeled vibrant coral reefs, hiked through ancient rainforests, stayed with Indigenous people in remote up-river communities. I'd been moved by the vastness of the desert sky, rugged mountaintop vistas and the peace of ancient temples at dawn. I'd also witnessed crushing poverty, dislocated cultures and landscapes ransacked by capitalism. All of this fueled my motivation to better understand the world; to grind through difficult chemistry labs, calculus tutorials and mountains of reading to understand how humanity had arrived at the place we now found ourselves in. It equipped me with the scientific tools I needed to try and do something about it, but what could I actually do to help the world?

After years of studying and work, I wanted to experience landscapes and cultures beyond my own backyard. I needed to take time to figure out what I wanted to do with my life; all I knew for sure is that I wanted to explore more of the world. I saved up all the money I earned from my tutoring job in the geography department until I eventually had enough to buy a round-the-world ticket. In 2001, we took off on a journey that would change the course of our lives. Captivated by my travel tales, my partner – a country boy who had never set foot outside of Australia – bravely journeyed with me across Hawai'i, Ecuador, Peru, Bolivia, Costa Rica, Nicaragua, Honduras, New York, Morocco, London, India, Nepal and Hong Kong until we ran out of money nearly a year later. I delighted seeing things from his fresh perspective; his enthusiasm for the unknown, his patience with setbacks, and his warmth befriending locals wherever we went. As the months slipped by, I also shared in his heartbreak as he witnessed suffering that sometimes made him retreat and weep. His response to the world made me feel less alone with my own grief – finally I had a companion willing to grapple with the complexity of life, the eternal challenge of balancing joy with sorrow. We became each other's light in the dark.

As we spent hours on rickety buses and overcrowded trains traveling across vast tracts of the developing world, an uneasy awareness started to dawn on me. Here we were – two Australians on an endless adventure – while the world around us was struggling to make ends meet. I'd seen poverty before, visiting relatives in Egypt and traveling throughout Southeast Asia, but my new scientific knowledge now provided me with deeper insight into what I was witnessing. In Latin America, on top of crippling poverty, people were still recovering from the devastation of the 1997–1998 El Niño event. When some of our luggage failed to arrive on one leg of our journey, we bought a replacement tent from a man who recounted the horror of Hurricane Mitch, the second-deadliest Atlantic hurricane on record, which caused over 11,000 deaths in the Central American countries of Honduras and Nicaragua in October 1998. Brackish water now caked everything they owned (including the tent we acquired from him); things hadn't recovered years on. People were still struggling to get back on their feet.

In Ecuador, we volunteered with a newly formed ecotourism company based in Quito looking to share their patch of the Amazon with tourists. In return for Spanish lessons and a cheap place to stay, we helped them develop their website and offered to test out one of their multi-day hikes. We visited regions where torrential rain and landslides had swept away entire villages. We battled knee-deep mud through cleared patches of rainforest now planted with cash crops, and met children in remote settlements who had never held a pen and paper in their hands. The uneasy alliance between economic development and the environment was clearly failing the most vulnerable – poor, working-class people and dislocated Indigenous communities were being left behind. It was hard to witness that level of poverty up close.

The reality of the cumulative impact of colonialism, capitalism and natural disasters was something I couldn't unsee. It became clear to me that climate change was going to wreak so much destruction in places that are barely above the poverty line. The people who have the most to lose, with meager means to protect themselves, will be hit hardest. As a daughter of parents from the developing world, I was taught very early on that life was not easy or fair. You couldn't rely on any handouts; you simply had to make the most of what you had. If you were one of the lucky ones, you got an education and managed to scrape together a better life for yourself; nothing was guaranteed except the need for hard work. If you did manage to succeed, your job was then to help the people around you to do the same.

My father was a classic example of a person who rose from humble beginnings to dedicate his life to helping others. He was one of seven children, born on the outskirts of Cairo in 1930 as the Great Depression took hold. His father went broke during the economic crash and then died when my dad was just six months old, robbing the family of their breadwinner, plunging them into poverty. While everyone had to do what they could to make ends meet, his older brothers insisted that he still went to school. He was bright and hardworking, so he exceled, going on to study medicine. The problem was, he still didn't have much money, so he often had to scale the security fence to sit his university exams because they charged students a fee to cover the costs of running exams at a public university during the late 1940s.

He slogged through his degree, and eventually went on to become a surgeon working in Egypt, Libya, Nigeria, Ireland, Scotland and, eventually, Australia. He worked as a doctor for six days a week until he was seventy-six, when advancing blindness eventually forced him to retire. He was loved by his patients and was a respected elder who provided counsel and care to generations of Middle Eastern migrants fleeing poverty or war. In his practice in southwestern Sydney he was known as "El Hakim," which means the "wise one." When he died in 2017, an overflow of mourners spilled out into the church courtyard, so great was his legacy.

The most enduring lesson my father imparted to everyone around him was captured by his motto: "education is the key to a better life." His own struggles taught him that knowledge is the great equalizer that allows people to overcome entrenched inequalities. When I was a child, listening to his stories made me feel like attending school was a sacred privilege; so even when I was sick, I strongly resisted missing a single day. So great were my tears, he had to try and persuade me to stay home with the reward of a one-dollar bill. Back then, that meant money to buy books, so eventually I'd give in.

As I traveled around the world, my dad's words took on new meaning. It became clear that I was leading a life of unimaginable privilege; I had been given an incredible opportunity to give something back to the world. The question was what to do with the cards I'd been dealt? The more I experienced, the clearer things became. In Peru, I managed to find an internet cafe and emailed my supportive lecturer back home: I wanted to do a PhD on El Niño – the largest source of climate variability on the planet, which was causing so much destruction to people and places everywhere I looked. I found it hard to engage in the vacuous conversations in backpacker hostels where wealthy people – almost exclusively from Europe and North America – boasted about how many countries they had "done," trying to outdo each other with the most exotic place they had visited.

Although I enjoyed the freedom of an endless holiday, ultimately I wanted to use the opportunities I had been given to help the world in some way. Over the coming months, I cobbled together a scholarship application, mailing it back to Australia from New Delhi with newfound clarity. When I returned home, months later, a letter was waiting for me – I'd been offered a scholarship to develop a global history of El Niño events. I felt like my life now had a meaningful direction – I could put my precious education to good use as an Australian to try and ease the suffering I saw in the world.

That's the long answer to how I came to be where I am today. A future career as a climate scientist is something I never would have predicted for that sensitive bookworm back in the mid-1980s. Decades on, working on an IPCC report felt like a global gathering of people just like me – people who care about the world; the ones who can't look away.

*   *   *

Holding the full scale of the planetary emergency in your mind – all at once – is an overwhelming experience, even for scientists. Until my work with the IPCC, it was hard to join the dots between the trends being observed where I live in Australia and events happening all over the world. The experience taught me so much, but perhaps the most important insight was that the people alive today will determine humanity's future.

Stop for just a moment to really let that sink in – averting planetary disaster is up to us.

It's a big thing to claim, so let me carefully explain why we are at such a fateful fork in the road.

Long-term climate records show us that carbon dioxide levels rise and fall in line with global temperature changes. What we do to reduce greenhouse gas emissions over the next decade is absolutely crucial for determining the kind of future we will experience in years and centuries to come. So much so that many experts call it the "critical decade," when fossil fuel use needs to peak then begin to rapidly decline. The IPCC's Working Group 2's report *Impacts, Adaptation and Vulnerability* very clearly states that the choices we make in the next decade will determine humanity's future. The key message from the report, which UN secretary-general António Guterres described as "an atlas of human suffering and a damning indictment of failed climate leadership," uses the most direct and urgent language I've ever seen in an IPCC Summary for Policymakers:

> The cumulative scientific evidence is unequivocal: Climate change is a threat to human well-being and planetary health. Any further delay in concerted anticipatory global action on adaptation and mitigation will miss a brief and rapidly closing window of opportunity to secure a liveable and sustainable future for all.

As Working Group 1 scientists presented in our volume of the report, at current emission rates, we only have around ten years before we reach 1.5°C of warming. We know that every fraction of a degree matters when it comes to climate change. Working Group 2 concluded that global warming of just over 1°C has already caused dangerous and widespread disruption to nature and human societies, affecting the lives of billions of people despite our efforts to adapt. The report clearly shows that we are currently failing to outpace escalating and compounding risks in many parts of the world, particularly in Africa, South Asia, Central and South America, the Arctic and small island nations.

The situation progressively worsens once we reach 1.5°C, when the adaptation limits of many ecosystems are reached. Beyond 2°C, adaptation is simply not possible in some low-lying coastal cities, small islands, deserts, mountains and polar regions. The Earth's history tells us that even small changes in temperature – less than 1°C due to natural factors alone – have resulted in major changes in the past. The monumental scale of human modifications to the planet is now well and truly dwarfing natural processes, meaning that the only way out of this mess is through our own actions. In fact, the IPCC identifies human-caused greenhouse gas emissions as the key factor that will determine the trajectory the Earth's climate takes. That is, the climate policies the governments of the world enact *right now* will seal our planet's fate.

Climate model simulations run with different greenhouse gas emissions are used to investigate a range of possible futures to help decision makers understand the consequences of our collective action. Individual models are based on mathematical representations of the physics, chemistry and biology of the atmosphere, land, oceans and ice sheets, run by the largest supercomputers in the world to generate future projections of the Earth's climate. Models are developed by independent research groups from all over the world, then their simulations are contributed to a freely accessible global database co-ordinated by the World Climate Research Programme's Coupled Model Intercomparison Project, known as CMIP, which is updated in line with IPCC assessment timelines. Phase 6 of this initiative, referred to as CMIP6, provides simulations run by the latest-generation models that underpin the IPCC's *Sixth Assessment Report*. CMIP6 consists of model "runs" from close to 100 climate models produced by more than fifty different modeling centers around the world. These simulations range from very low emissions scenarios that assume strong climate policy implementation and improved equality in overall socioeconomic development goals, to very high emissions scenarios in the absence of additional climate policies and the adoption of resource-intensive, fossil fuel–based lifestyles around the world. Researchers then translate these socioeconomic conditions into estimates of future energy use and greenhouse gas emissions to provide a range of climate change projections.

In CMIP6, a range of possible climate futures are investigated using nine low-, medium- and high-emissions scenarios, referred to as Shared Socioeconomic Pathways (SSPs). These are based on five broad socioeconomic scenarios that are intended to represent a range of plausible futures. They include: a world of sustainability-focused growth and development goals that emphasize equality (SSP1); a "middle of the road" scenario where trends broadly follow their historical patterns (SSP2); a fragmented world of "resurgent nationalism" that focuses on regional interests at the expense of global goals (SSP3); a world of increasing inequality and continued investment in fossil fuels (SSP4); and a world of rapid and unconstrained economic growth based on abundant fossil fuel use (SSP5). These five baseline scenarios are designed to investigate how things would look in the absence of climate policy and allow researchers to examine underlying socioeconomic conditions that affect the implementation of energy policy and climate change adaptation strategies. These scenarios cover a broader range of greenhouse gas and air pollution futures than assessed in previous IPCC reports, providing a very extensive look at a range of possible pathways humanity might take. It's important to realize that the IPCC does not assign any probabilities associated with the likelihood of these scenarios – the concentration of greenhouse gases in the atmosphere is something determined by the collective climate policies that governments all over the world choose to implement or not. That is, the path we take is literally in our hands.

SSP1 and SSP5 are based on optimistic trends for human development, with improvements in education and health, rapid economic growth, and well-functioning institutions. The point of difference between the two scenarios is that SSP5 assumes this development will be driven by an energy-intensive, fossil fuel–based economy, while in SSP1 there is a shift towards sustainable practices. SSP3 and SSP4 are more pessimistic in their future economic and social development outcomes, with little investment in education or health in poorer countries coupled with a fast-growing population and increasing inequalities. SSP2 was designed to be a "middle of the road" scenario where historical patterns of development are continued throughout the twenty-first century. Together these scenarios are the most comprehensive ever considered by an IPCC assessment.

The core set of five SSP scenarios featured in the IPCC's *Sixth Assessment Report* are SSP1–1.9, SSP1–2.6, SSP2–4.5, SSP3–7.0 and SSP5–8.5, where the numbers refer to the amount of extra energy in the atmosphere due to heat-trapping greenhouse gas emissions by the end of the century, measured relative to pre-industrial levels (1850–1900) in watts per square meter. This is known as radiative forcing, the amount by which the Earth's energy budget is out of balance: the higher the number, the more greenhouse gases are trapping heat in the atmosphere instead of being reflected back into space, causing the planet to warm.

In terms of the low-emissions scenarios that are consistent with achieving the targets set out in the Paris Agreement, SSP1–1.9 holds warming to approximately 1.5°C above pre-industrial levels by 2100, with net-zero emissions achieved by the middle of the century. With

SSP1–2.6, global warming remains below 2°C, with net-zero emissions achieved in the second half of the century. The thing to realize about these low-emissions pathways is that they both include varying levels of what are termed "negative emissions." That is, large-scale carbon dioxide removal from the atmosphere over a sustained period through unproven and as yet unavailable technologies such as carbon capture and storage, as well as ready-to-implement nature-based solutions such as planting trees and restoring coastal and ocean ecosystems. As you'll see later on, relying on unproven technology to materialize and magically save the day is a distraction we simply cannot afford.

Right now, global emissions pledges under the Paris Agreement – known as Nationally Determined Contributions, or NDCs – are not on track to achieve either of these low-emissions scenarios. That is, we are not on track to stabilize warming below 2°C. Instead, if current NDC pledges for 2030 are implemented in tangible policies, our trajectory is approximately in line with the intermediate-emissions pathway of SSP2–4.5. This can be thought of as a "no additional climate policy" scenario that results in very likely global temperature increases of 2.1–3.5°C, with a central estimate of 2.7°C of global warming by the end of the century. I think of the SSP2–4.5 pathway as a best-case scenario based on where things are right now. But given that global emissions are still rising despite policy pledges, and the extreme events being experienced right now are beyond what climate models are currently able to predict, only considering a "middle of the road" scenario feels overly optimistic to me. Leaving aside the very real barriers to implementing strong climate policy during an ongoing pandemic and increasing political instability, these intermediate projections may underestimate the true extent of the physical changes we may experience, so the scientific reality of our true position may catch society off guard.

That brings us to the sobering set of high-emissions projections, SSP3–7.0 and SSP5–8.5, which would result from no or no additional climate policies and a failure to substantially reduce greenhouse gas emissions before 2050. You can think of these as our worst-case scenarios. SSP3-7.0 lies between the intermediate SSP2-4.5 pathway and the lower end of the SSP5-8.5 pathway based on unmitigated fossil fuel use. Unlike other scenarios, the SSP3-7.0 pathway factors in particularly high non-carbon dioxide emissions like methane, nitrous oxide and sulfur dioxide, land use changes resulting from decreased forest cover, and high concentrations of aerosols associated with increased air pollution. In this case, we see global warming in the range of 2.8–4.6°C, with a central estimate of 3.6°C by 2100. I consider SSP3-7.0 a plausible scenario based on *currently implemented policies* – not well-intentioned promises – and uncertainties in climate modeling such as carbon cycle feedbacks, which are expected to amplify warming, and the higher sensitivity of temperature to increases in carbon dioxide than previously thought. But more on that later.

Finally, we have the nightmarish SSP5-8.5 "no policy" pathway, which can be considered the most extreme end of the worst-case scenario, involving high population growth and continued reliance on fossil fuel-based development. This very high emissions scenario is a warning of what could happen if we continue to depend on fossil fuels, with the models indicating a temperature increase of 3.3–5.7°C by 2100, with a best estimate of 4.4°C. While some experts believe that major improvements in renewable energy technology and the overly high sensitivity of some climate models make this outcome unlikely, it is still considered possible to reach very high levels of greenhouse gas concentrations under lower emission trajectories due to carbon cycle feedbacks that may accelerate warming. Consequently, the IPCC report presents the SSP5-8.5 pathway as a useful scenario for examining extreme warming arising from unexpected non-linear processes, such as abrupt permafrost thaw, which are currently inadequately captured by climate models. It also serves as a stark warning of the consequences of weak climate policy and the absence of readily deployable carbon capture and storage technology to remove large volumes of carbon dioxide from the atmosphere.

These are confronting numbers to take in, especially when you consider that the IPCC highlights that in all scenarios considered in the report, the central estimate of crossing the 1.5°C global warming level lies in the early 2030s (with the exception of SSP5-8.5, which would see the breach occur earlier). Global warming of 2°C above pre-industrial levels is extremely likely to be exceeded during the twenty-first century under the three scenarios that do not decline greenhouse gas emissions before mid-century (SSP2-4.5, SSP3-7.0, SSP5-8.5). The report concludes by saying that sustained global warming levels of more than 2.5°C higher than pre-industrial levels have not occurred in over 3 million years, when major elements of the climate system – like the extent of ice sheets and position of coastlines – were very different from today. It's a very blunt warning that unless we drastically change course and cut emissions in half by 2030, and reach net-zero emissions no later than 2050, we are on track for catastrophic levels of warming that will profoundly alter all life on Earth.

*   *   *

One of the most concerning features of the new CMIP6 projections is the number of models that show very high levels of warming relative to increases in carbon dioxide levels. What really worries scientists is the concept of how sensitive the Earth is to establishing its new equilibrium once its balance has been disturbed. The most common way to measure and investigate the effects of global warming is known as "equilibrium climate sensitivity" (ECS), defined as the full amount of global surface warming that will eventually occur in response to a doubling of atmospheric carbon dioxide concentrations compared to pre-industrial levels. Not knowing the true sensitivity of the Earth's temperature to increased greenhouse gases is what causes the largest uncertainties in global temperature projections (aside from the uncertainty of which future emissions pathway humanity will take). That's why equilibrium climate sensitivity is sometimes referred to as the "holy grail" of climate science because it helps quantify the risks posed to human society as the planet continues to warm.

We know that carbon dioxide concentrations have risen from pre-industrial levels of 280 ppm to approximately 416 ppm in 2021 – increasing at a rate that is at least ten times faster than at any other time during the past 800,000 years. Note that when scientists refer to greenhouse gas concentrations like 280 ppm, it means there are 280 molecules of a given gas per million molecules of dry air. It is expected that 2022 will be the first year when carbon dioxide levels will be 50 percent above pre-industrial levels. Without major reductions in greenhouse gas emissions, we are likely to reach 560 ppm – the doubling of pre-industrial levels – sometime between 2060 and 2080.

When the IPCC's *Fifth Assessment Report* was published in 2013, it estimated that such a doubling of carbon dioxide was likely to produce warming within the range of 1.5°C to 4.5°C as the Earth reaches its new equilibrium. Revised estimates calculated from four lines of evidence, including the latest global climate models used in the IPCC's *Sixth Assessment Report*, are now higher than before. Improved knowledge of climate feedbacks (particularly those related to cloud processes), conditions experienced during past climates, and improved quantification of observed changes in the Earth's energy imbalance, have led to improved estimates of equilibrium climate sensitivity. The new numbers suggest that a doubling of carbon dioxide may in fact produce between 2°C and 5°C of warming, with a central estimate of 3°C. In a nutshell, the Earth is more sensitive to changes in carbon dioxide levels than we once thought, meaning that we might have underestimated the amount of warming now baked into the system over coming centuries. The IPCC starkly warns that it is "currently not possible to rule out ECS [equilibrium climate sensitivity] values above 5°C."

Reflecting on the new results, two of Australia's leading climate scientists, Michael Grose and Julie Arblaster, noted in an article in the online news site *The Conversation* that "the new values are a worrying possibility that no one wants, but one we must still grapple with." They quoted the authors of another recent climate study, who said, "what scares us is not that the models' ECS [equilibrium climate sensitivity] is wrong . . . but that it might be right."

When the initial results were first released at a climate modeling workshop in March 2019, a flurry of panicked emails from my IPCC colleagues flooded my inbox. What if the models are right? Has the Earth already crossed some kind of tipping point? Could we be experiencing abrupt climate change right now? When many of the most advanced models in the world are independently reproducing the same disturbing results, it's hard not to worry. Since then, over a third of the latest models assessed in the IPCC's *Sixth Assessment Report* – including simulations run by leading research centers in the United States, the United Kingdom, Australia, Canada and France – show climate sensitivity of 4.5°C or warmer. The report warns that it is not possible to rule out eventual, long-term global warming above 5°C based on currently available science. Instead, the IPCC explains that models with high equilibrium climate sensitivity are useful for understanding the risks posed by extreme warming scenarios that are still very much on the table. Perhaps seeing the reality of future scenarios that would obliterate life as we know it will finally help jolt governments out of their complacency, compelling them to deploy an emergency response to cut global greenhouse gas emissions to net zero no later than 2050.

When the UN's Paris Agreement was adopted in December 2015, it defined a specific goal: to keep global warming to well below 2°C and as close as possible to 1.5°C above pre-industrial levels to avoid dangerous levels of climate change. Examining the Earth's climatic past tells us that even between 1.5 and 2°C of warming will trigger huge reconfigurations of the planet's climate system. At sustained warming levels between 2°C and 3°C, there is evidence that the Greenland and West Antarctic ice sheets will be lost almost completely and irreversibly over thousands of years. As we'll see a little later on, the planet becomes unrecognizable between 3°C and 4°C. In these high-warming scenarios, the Earth as we once knew it will no longer exist.

While there is no going back from some changes, what we do right now will reduce the scale and magnitude of the disaster. We have reached a fateful fork in the road. How we choose to respond to rising greenhouse gas emissions over the next five to ten years will determine our future climate conditions and the fate of human societies for thousands of years to come.

The question is, can we muster the best of our humanity in time?

*   *   *

At the ends of the Earth, the elements are sculpted into otherworldly landscapes that have captivated humans for centuries. Creaking blue glaciers splintering into frozen seas, icebergs chiseled by ferocious polar winds, powdery plains concealing mythical white creatures. Dark, beady eyes give away the presence of a startling Arctic fox, its thick coat of pristine white emerging seamlessly from the labyrinth of its underground den. In the waters of the North Pole, narwhals – their enormous spiral tusks extending 1.5–3 meters from their skulls – earn their title of "unicorns of the sea." Long ago, their horns were considered to have magic powers, like neutralizing poison and curing melancholia, allowing Vikings to trade them for many times worth their weight in gold.

Of all the Arctic's unique wildlife, the most iconic is the polar bear, a giant maritime beast that lives on the sea ice of the Arctic Circle. Adult males can weigh more than 600 kilograms with huge, 30-centimeter paws equipped with formidable 5-centimeter claws to rip apart prey. The layer of frozen seawater that floats on the ocean surface freezes and melts in turn with the seasons, allowing the bears to roam for long distances to hunt seals that live in the open sea fringing the icy coasts of Canada, Alaska, Russia, Norway and Greenland. As a child of the 1980s, one of my first inklings that all was not well in the world came about from learning that polar bears were drowning out at sea because global warming was melting sea ice, stranding them away from their homes. They became a visible symbol of an endangered world; their plight shone a light on the fragility of life on a planet being rapidly warmed by humans.

Unlike Antarctica, which is a vast ice-covered continent surrounded by ocean, the Arctic is a sea covered with surface ice surrounded by land. Polar opposites, if you will. During the winter months, the open ocean surface of the Arctic solidifies into huge sheets of ice that are typically 2 meters thick, but can accumulate up to 5 meters in remote areas north of Greenland that stay frozen during the summer melt season. The presence of sea ice is critically important as its bright surface reflects sunlight back into space instead of absorbing heat and warming the ocean. But as temperatures increase, there are fewer reflective surfaces to deflect incoming heat, resulting in more solar energy being absorbed by the darker ocean. This causes temperatures to rise even further, resulting in the phenomenon known as polar amplification that we covered earlier.

So although future projections in average global temperature are shocking enough, it's the changes in the polar regions – the frozen areas of the world that keep our climate stable – that really keep me up at night. It's alarmingly clear from current observations that the Arctic is already undergoing rapid changes that will substantially alter the Earth's climate for centuries to come – but by exactly how much? It's a difficult question to answer because carbon and methane emissions from permafrost melt and Arctic wildfires are not fully accounted for by most models, including those used in the latest IPCC report. This is because of limited observations available for these remote and underground regions, and the incredible difficulty of simulating complex processes like abrupt thawing of permafrost and changes in vegetation in models. This makes the climate change projections for the Arctic Ocean some of the most confronting results of the entire IPCC assessment report, as they are likely to be an underestimate.

Under a "middle of the road" emissions scenario (SSP2–4.5), the latest IPCC models indicate average warming in the Arctic of 9.2°C above pre-industrial conditions by the end of the century, with a very likely range of 5.4–14.4°C. Unbelievable, I know. Average temperature increases of 4.7°C above pre-industrial levels are projected in the near term (2021–2040). If that's not bad enough, the worst-case, high-emissions scenario (SSP5–8.5) simulates an average temperature increase of 13.9°C, with the warmest models in the very likely range indicating an unimaginable warming of 19.9°C in just eighty years. Under high emissions, on average the Arctic Ocean could warm by 5.1°C above pre-industrial levels over the next twenty years. When I showed these results to a colleague, his response was, "This makes me want to lie down and cry." Me too.

Let me explain why it's hard to come to terms with these projections for the Arctic. Actual observations of recent temperatures in the Arctic are already exceptionally pronounced – warming well above the global average. Estimates from satellite records show a rate of annual warming over sea ice–covered regions of 0.47°C per decade between 1981 and 2012, with an astounding rate of 0.77°C per decade over Greenland. But it's the warming trend over the Barents and Kara

Seas, located off the northern coasts of Russia and Norway, that is truly shocking – a staggering 2.5°C per decade. To make things worse, permafrost warming and thawing have been widespread in the Arctic since the 1980s, but we are yet to fully estimate the impact of these changes. The thing to remember here is that these are not model simulations of the future; these are conditions that have *already been observed*, making the projections seem plausible rather than existing in the realm of science fiction.

Under high-emissions scenarios, the Arctic Ocean is projected to become practically ice-free in late summer by the end of the twenty-first century, which will further amplify warming through a range of feedback processes, accelerating the destabilization of the Earth's climate. Even under low-emissions scenarios, summer sea ice will disappear nearly every September starting at around 1.7°C global warming. The IPCC report warns that continued warming of the region will lead to further reductions in Northern Hemisphere snow cover and near-surface permafrost. Even if global temperatures are stabilized immediately, glaciers will continue to lose mass for at least several decades before a new equilibrium is reached. The 2021 *State of the Cryosphere* report (with its pointed subtitle: "We cannot negotiate with the melting point of ice") highlights that steep declines will continue even under low-emissions scenarios, with no tropical glaciers and few mid-latitude glaciers outside the Himalayas by 2200 if 2°C of global warming is reached. Once 2°C is passed, the only glaciers of any substantial size will be limited to the polar regions and the highest mountain ranges like the Himalayas by 2300. And even in these regions, glaciers could shrink to half or a third of their current size. Recovery from that kind of warming would take mountain glaciers over a thousand years, with ice sheets not growing back for more than 10,000 years.

What really concerns me is that the latest IPCC assessment found that at sustained warming levels between 2°C and 3°C, there is evidence to suggest that the Greenland and West Antarctic ice sheets will be lost almost completely and irreversibly over thousands of years. With 2°C of warming, committed global mean sea level rise will be about 2–6 meters over 2000 years. Between 3°C and 5°C, the Arctic Ocean will be practically ice-free throughout September in most years, with models indicating the near-complete loss of the Greenland ice sheet and complete loss of the West Antarctic ice sheet. This drastic loss of ice will result in a radically transformed planet. Ice sheets contain enormous quantities of frozen water, so when they melt, they unleash large volumes of water that will raise global sea level.

If the Greenland ice sheet melts, there is the potential to increase sea level by around 7 meters, with West Antarctica adding an additional 4 meters. The IPCC estimates that global sea level will increase by 12–16 meters over 2000 years under a peak warming of 4°C, and an inconceivable 19–22 meters with 5°C. The loss of the world's ice sheets will also weaken ocean–atmosphere circulation patterns as the temperature gradient between the equator and the poles becomes less pronounced, and ocean currents are disrupted by the influx of freshwater altering ocean salinity. Under such high levels of warming, nearly all glaciers in Central Europe, the Caucasus, Western Canada, the United States, North Asia, Scandinavia and New Zealand will also likely disappear, reinforcing a dangerous warming trend that will reconfigure planetary conditions.

It's important to realize that human civilization has adapted to sea levels that have remained stable over the past 7000 years. Such radical changes in our coastlines – even if they take place over such a long time period – will profoundly alter the face of our planet. Close to 1 billion people currently live in the low-lying coastal zone (located less than 10 meters above sea level), including the cities of Bangkok, New York, Shanghai, London, Calcutta and Sydney. As of 2020, it is estimated that over a quarter of a billion people live on land less than 2 meters above mean sea level, which is the area most exposed to future sea level rise. That number could increase to 410 million by 2100, with nearly three quarters of this land located in the tropics. The most vulnerable region disproportionally lies in tropical Asia, taking in countries like Indonesia, Bangladesh and Thailand. In the Ganges-Brahmaputra-Meghna delta in the Bay of Bengal, where between 20 and 60 percent of land is already flooded every year, further sea level rise will expose tens of millions of people to increased flood risk. In Bangladesh alone, land below 2 meters currently houses a population of 18.1 million. With a sea level rise of just 1 meter, 4.9 million people would be below sea level. And this is just in one country. And that's just for average sea level rise, not extreme sea level events associated with severe storms. As we will soon see, relocating a huge proportion of the human population displaced by rising seas and inundation from repeated storm surges will be a colossal challenge – if not impossible to cope with.

The thing to remember here is that these projections are not in the realm of science fiction; warming already in train will alter low-lying areas of the world forever. How quickly we act now will determine how many people we can save.

4

Gradually, then suddenly

LET ME APOLOGIZE FOR BEING the bearer of bad news. I know the science is a lot to take in. It's terrifying and overwhelming. But just like dealing with a medical emergency, it's important to know exactly how bad things are so you know what realistic treatment options are available. So please know that I am just trying to convey the gory details as simply as I can, before you can decide what you want to do about it.

The latest IPCC assessment is best described by UN Secretary-General António Guterres, who said that the report is a "code red for humanity." The world has been coasting along, oblivious to the escalating scale of the climate crisis. But now that our climate model projections are horrifically springing to life as each season passes, people no longer need to use their imaginations to picture climate change – it's here, right now, part of life for every single person on the planet. The evening news now looks like a disaster movie. It's impossible to stress how important it is to slam on our emergency brakes to avoid screeching past the point of no return. Any further delay in reducing global greenhouse gas emissions will cost us our planetary stability, unleashing untold destruction for countless generations. We must move quickly.

One of the most significant features of the IPCC's *Sixth Assessment Report* is its detailed coverage of abrupt and irreversible climate change. This is direct acknowledgment from the scientific community that there can be sudden, unexpected changes in the Earth's climate that can unfold very quickly, particularly under higher levels of global warming. Once these switches have been flicked, they have the potential to trigger severe impacts for human and natural systems, like major changes in the water cycle, temperature extremes, and the large-scale dieback of ecosystems. How far away we are from these critical transitions is an area of active research. Understandably, abrupt climate change is a topic of intense public interest that catches the attention of the media. But unfortunately it is also an area that is often poorly understood by people outside of the scientific community, so it is worth covering the basics so you are aware of where the current science stands.

The IPCC defines abrupt change as global- or regional-scale climate change that occurs substantially faster than the typical rate of change in its recorded history. In other words, unexpected, non-linear jumps in the climate are possible. In fact, they've happened before. The Earth's history contains examples of abrupt climate change that can happen in just a handful of decades or even years. Obviously it's a timescale that is short enough to challenge the capacity of human societies and ecosystems to adapt, so we need to understand the risks posed by these "low likelihood, high impact" scenarios. In many cases, abrupt changes arise from gradual shifts in one component of the Earth system like the disintegration of ice sheets, permafrost thaw, slowdown of ocean circulation or the dieback of tropical and boreal forests. Once a critical threshold is breached, the system becomes unstable and reorganizes dramatically, and sometimes irreversibly. A change in one part of the Earth can sometimes trigger a cascade of impacts in other components of the system – like a giant set of dominoes rippling across the planet.

Interactions between the frozen areas of the Earth and the ocean have produced some of the most dramatic events in our geologic past.

Pulses of freshwater from melting glaciers and catastrophic iceberg discharges have altered the circulation of the ocean in the past, to such a degree that the tropical rain belt shifted southwards, significantly disrupting the behavior of regional monsoons. This plunged some areas into prolonged periods of aridity known as "megadroughts," while other regions became inundated by intensified monsoon rains. These changes resulted in dramatically altered conditions, sometimes causing societal and ecological collapses around the world; the repeated failure of floods on the Nile River, for example, contributed to the political shift from the Old to New Kingdom in Egypt over 3500 years ago, and intense drought led to the collapse of the Khmer Empire at Angkor in Cambodia between AD 1300 and 1500.

The geologic record contains examples of abrupt climate change that can happen in just a handful of decades or even years. During the last ice age cycle, ice core records reveal that the North Atlantic experienced a series of dramatic climatic fluctuations when conditions alternated between marked warming around Greenland followed by a gradual cooling. These are referred to as Dansgaard–Oeschger events, named after scientists Willi Dansgaard and Hans Oeschger who first reported the feature from Greenland ice cores during the 1990s. About twenty-five of these abrupt warmings have occurred over the past 120,000 years, with temperatures in some regions increasing as much as 5–16°C within a few decades. The length of these warmings varies from a century to several thousand years. The most striking example is an abrupt event that occurred around 14,700 years ago when Greenland ice cores indicate a 10°C temperature increase within three years. The ice core records also reveal that during these abrupt warmings, methane entered the atmosphere 50 percent faster than it had previously, as wetlands re-emerged as the Arctic thawed out. The evidence shows that although these events were initiated in the North Atlantic region, they triggered major changes right across the planet.

Scientists are now in the position of trying to figure out what to expect if these sudden changes occur as the world reaches higher levels of global warming. Abrupt change often arises from self-reinforcing (positive) feedbacks that cause the climate system to become unstable. When this happens, a "tipping point" can be crossed, causing a rapid shift from one state to another. A tipping point is when a part of the Earth system reaches a critical threshold when a small change leads to significantly larger changes in other major components, namely major ecological communities (biomes) such as the Amazon rainforest, boreal forests and coral reefs, polar ice masses like Arctic sea ice and the West Antarctic ice sheet, and major circulation systems in the atmosphere or ocean, including changes in the North Atlantic Ocean. It's a classic case of a small change making a big difference: the proverbial straw that broke the camel's back.

Once a system has crossed into a different state, it stays there for a very long time, sometimes even permanently. For example, part of the reason that Greenland has an ice sheet today is that it has had an ice sheet there for hundreds of thousands of years. If the Greenland ice sheet passes a tipping point this century that leads to its complete disintegration, reducing emissions and lowering global temperatures to pre-industrial levels would not bring it back anytime soon. It would probably require another ice age to achieve that, and we know these operate on 100,000-year cycles, well outside the timeframe that human societies can deal with. It's also important to realize that these conditions were initiated when carbon dioxide concentrations were well below 300 ppm – and we are well over 400 ppm right now, with no sign of slowing down. Experts have analyzed past warm periods in the Earth's history, concluding that the next ice age is unlikely to happen anytime within the next 50,000 years.

The other thing to keep in mind about tipping points is that they don't operate in isolation. Many are linked, so the destabilization of one element may influence the triggering of another, known as a "tipping cascade." The best-known example is the melting of Arctic sea ice intensifying regional warming because darker ocean water absorbs more solar radiation than white, reflective sea ice. The ice acts like a giant mirror, reflecting sunlight back into space, but once it's gone, the ocean begins to soak up more heat. This increases the rate of melting around the Greenland ice sheet, which comes into contact with warmer seawater. As the ice sheet recedes, the amount of meltwater flowing into the North Atlantic increases, altering one of the world's most important areas of ocean circulation that distributes heat from the equator to polar regions, and back again. This can cause a weakening or shutdown of the ocean circulation in this region, known as the Atlantic Meridional Overturning Circulation, or AMOC, causing major disruptions to the climate of northern Europe and the tropical rainfall belt, which generates regional monsoons that currently sustain billions of people around the world. The collapse of the AMOC was the inspiration for the movie *The Day After Tomorrow*, when the Northern Hemisphere is plunged into an ice age as the ocean current responsible for bringing warmth from the tropics switches off.

The AMOC forms part of a wider network of global ocean circulation patterns that transport heat around the world. It is driven by deep water formation, which is influenced by conditions at the ocean surface. Usually warm, salty water in the upper layers of the ocean travels towards the poles, where it begins to cool, making the water mass denser. Once it becomes heavy enough, this denser water then sinks and flows towards the equator at much greater depths. Climate change is now affecting this process by diluting salty seawater with freshwater and warming the surface layers of the ocean. This dilution happens through increased rainfall and the melting of continental ice, particularly around the Greenland ice sheet. As this freshwater makes water lighter and less able to sink, the overturning circulation begins to slow down, altering the behavior of the ocean and the atmosphere above it.

There is emerging evidence from instrumental ocean measurements and marine sedimentary records that suggests that the AMOC is starting to weaken. An initial slowdown began during the late nineteenth century, followed by a second, more rapid decline since the mid-twentieth century. While these findings are in line with theoretical expectations indicated by climate models, there are still insufficient records to quantify the magnitude of the weakening because direct sea surface temperature observations only extend back to the 1980s, so are too short to confirm that these findings are within the natural range of climate variability for the region. While longer sedimentary records used to investigate past changes in the AMOC show that the ocean circulation system is currently in its weakest state in more than 1000 years, more studies are needed to support a definitive conclusion. As a result, the latest IPCC report concludes that while the AMOC is very likely to weaken over the twenty-first century under all emission scenarios, confidence in the magnitude of the overall decline is low. Like the melting permafrost, it's a case of watch this space.

Because thermohaline changes associated with the shutdown of the AMOC is the mechanism identified as driving many of the abrupt changes seen in the Earth's history, it has been the intense focus of research in recent years. During the warming that followed the last ice age, there was a prominent slowdown of the AMOC during a period known as the Younger Dryas, around 12,800–11,700 years ago, which caused worldwide changes in rainfall patterns. The tropical rain belt migrated south, leading to a weakening of the African and Asian monsoons and the strengthening of wet seasons in South America and Australasia. These patterns are relevant for future projections as the IPCC considers it very likely that the AMOC will weaken during the twenty-first century in response to continued warming. Some model studies show that if the AMOC crosses a tipping point, a rapid weakening could occur over a few decades, followed by a slower decline that may play out over hundreds of years.

Based on current evidence, there is only medium confidence that there will not be an abrupt collapse of the AMOC before 2100, with the IPCC stating that the possibility of collapse beyond the twenty-first century under high levels of sustained warming "cannot be ruled out." But should it occur, it is very likely that there would be major impacts on the global water cycle, influencing half of the world's population that relies on monsoon rains. It would also see widespread cooling in parts of the Northern Hemisphere, especially around Western Europe and the east coast of North America, where average temperatures could cool by more than 3°C under a high global warming scenario. This would have major impacts on agriculture in places like Great Britain, where the extent of arable farmland could decline from 32 percent to just 7 percent from the regional drying effects of an AMOC collapse. As a result, the IPCC considers this a "low likelihood, high impact" scenario that is more likely under substantial warming in the order of 3-4°C above pre-industrial levels – an outcome that is still possible under the higher end of the IPCC's intermediate-emissions scenario that we are currently tracking. But the good news here is that, based on current evidence, a complete collapse of the AMOC is unlikely to happen over the next few decades. I know it doesn't sound too comforting, but at least it's something.

Another tipping point of major concern is the dieback of the Amazon – the world's largest tropical rainforest – through deforestation or conversion into drier savanna or open woodland, and ongoing climate change. The rainforest spans nine countries throughout South America; it is twice the size of India and bigger than Australia, so is a vital area for regulating the Earth's climate and the region's water cycle. Because trees are critically important for the removal of carbon dioxide from the atmosphere, any reduction in the ability of the Amazon to act as a huge "carbon sink" has major implications for future global warming. As humans convert rainforests into areas for beef and soy production, the internal recycling of moisture through evapotranspiration – the combined process of evaporation from land and the release of water from leaves – is reduced, drying out the rainforest. This is being exacerbated by regional warming, which is intensifying the dry season, leaving the rainforest susceptible to fire and longer periods of drought, which in turn releases vast amounts of carbon dioxide when vegetation burns. These disturbances further reduce the overall area of rainforest and evapotranspiration rates, reinforcing the vicious cycle of degradation that is likely to see this ecosystem cross a tipping point into a dry state this century. There is only so much drying the Amazon can tolerate before the rainforest stops being able to support itself.

A recent study by a large international research team, led by Brazilian scientist Luciana Gatti, reported that the southern and eastern regions of the Amazon are now releasing more carbon dioxide than they store as deforestation and fires become more widespread. As more trees die, the Amazon's ability to absorb carbon from the atmosphere is declining, further accelerating global warming. The rainforests in the eastern Amazon have already warmed by as much as about 0.6°C per decade during the dry season over the past forty years – more than three times the global average – heralding that major changes are now underway. Since the 1960s, it is estimated that around 20 percent of total Amazon forest cover has been lost as a result of deforestation and fires, with that figure expected to rise as high as 40 percent by 2050. The overall pattern of deforestation, warmer and drier dry seasons, drought stress and wildfires now seriously threatens the ability of these tropical forests to lock up large amounts of carbon dioxide in the future.

Given that photosynthesis is estimated to have absorbed around 25 percent of all carbon emissions from the burning of fossil fuel since

1960, any decline in the ability of the world's largest intact tropical forest to absorb carbon dioxide is a huge cause for concern. The IPCC concludes that continued deforestation, combined with a warming climate, raises the probability that the Amazon will cross a tipping point into a dry state during the twenty-first century. Limitations in modeling vegetation dynamics make it hard to know exactly when this might happen. Approximately 17 percent of the Amazon rainforest has already been destroyed, with deforestation rates rising again as far-right Brazilian president Jair Bolsonaro – sometimes referred to as "the [Donald] Trump of the Tropics" – has been implementing policies that encourage the clearing of the Amazon to make way for cattle ranches and cash crops, while rolling back protections for Indigenous people and undermining environmental conservation in the region. Some experts believe that the Amazon tipping point could occur when 20–25 percent of the rainforest has been felled, which could happen within fifteen to twenty years at current rates of deforestation. New research suggests that once tipped, the Amazon rainforest could shift into a dry, savanna state within fifty years, in line with process-based models and expert judgment. It's a terrifying prospect that may already be appearing on the horizon.

To make matters worse, scientists suggest that the interconnected nature of these processes could lead to a worldwide "tipping cascade," essentially setting into motion a global domino effect where humans may no longer be able to influence the climate trajectory. The good news is that there is limited evidence of abrupt climate change on a global scale up to 2°C of global warming. Although increases in global ocean heat content will continue until at least 2300 even under low-emissions scenarios because of the slow circulation of the deep ocean, current evidence does not indicate abrupt changes in the carbon cycle below 2°C. But at the regional scale, the IPCC states that abrupt response and tipping points that can have severe impacts "cannot be excluded."

The latest assessment concludes that risks associated with large-scale single events like ice sheet disintegration, biodiversity loss and changes in ocean ecosystem functioning, increase from "high risk" between 1.5 and 2.5°C of warming, to "very high risk" between 2.5 and 4.0°C.

Although the probability of crossing thresholds that trigger abrupt changes increases with higher levels of warming, a recent study by Dutch scientist Sybren Drijfhout found that eighteen out of thirty-seven abrupt regional shifts occurred in model simulations with global warming less than 2°C, the upper bound of the Paris Agreement's target designed to limit dangerous climate change. The results suggest that it is likely that the Earth will experience sharp regional transitions even under moderate levels of warming, with the collapse of some vulnerable ecosystems like the Amazon rainforest and tropical coral reefs occurring over human timescales of years and decades. In Australia, we have seen at least 50 percent of the Great Barrier Reef die off since 2016, so the reality of how quickly things can play out is something scientists understand all too well.

Now you know why I'm writing this book.

*   *   *

Looking back into the Earth's past is really helpful for understanding the type of conditions we might face if or when regional or planetary tipping points are crossed. It provides us with a way of imagining what life might be like if we shift into a climate that is radically different from today. One of the greatest strengths of the latest IPCC report is its use of geologic observations of past climates to provide context for the conditions we are currently experiencing. This field of science is known as palaeoclimatology: "palaeo" means old or ancient, and "climatology" is the study of accumulated weather over past seasons and years. By studying natural climate records like ice cores, marine sediments, corals and tree rings, we are teleported centuries back in time, providing scientists an unparalleled perspective on recently observed climate variability and extremes. Of particular interest are past warm periods, when carbon dioxide levels, temperature and sea level were higher than they are today.

The ancient warm period that has received the most attention is called the Pliocene, some 5.3–2.6 million years ago, because it could be the closest match for where we are headed. Around 3 million years ago, carbon dioxide levels were 360–420 ppm, similar to present-day levels of 416 ppm. During this period, the Earth experienced prolonged warming when temperatures were 2.5°C–4°C higher than pre-industrial times. Ocean temperatures in the high northern latitudes were up to 7°C warmer, and conditions were warm enough to support boreal forests instead of Arctic tundra. Farther south in the tropical Pacific, a weakened temperature contrast between the equator and the poles resulted in a "permanent El Niño" state. There was an intensification of monsoons in north Africa, Asia and northern Australia in this warmer world. This resulted in an expansion of tropical savannas and woodlands in Africa and Australia where deserts are found today. Geologic evidence and ice sheet modeling indicate that the Greenland and Antarctic ice sheets were much smaller than present, resulting in sea level rise of 5–25 meters around 3.3 million years ago. Our modern coastlines would be so far underwater, drowning many of the great cities of the world.

The fact that Pliocene carbon dioxide concentrations were similar to today's, while global temperatures and sea levels were significantly higher, allows scientist to analyze the sensitivity of the climate system independently of models. That is, we can examine the difference between an Earth system that has fully adjusted to changes in natural drivers in the deep past, and current conditions where greenhouse gas concentrations, temperature, and sea level rise are still increasing.

Insights from such studies may illuminate what sort of long-term changes might inevitably play out as the carbon dioxide levels we've already reached alter ice sheets over coming centuries. For example, the IPCC estimates that global sea level rise by 2300 ranges from 0.3 meters to 3 meters above pre-industrial levels under a low-emissions scenario, to as much as 16 meters higher following a very high emissions scenario that includes accelerating disintegration of polar ice sheets. Climate modelers use geologic records of past warm periods to help ground-truth their simulations with actual observations from the past. They are also a stark reminder that the long-term adjustment to present day carbon dioxide levels has only just begun; our world will be radically transformed for centuries to come. The IPCC report mentions that the mid-Pliocene warm period is a good analogue for the changes that might be expected from the SSP2-4.5 intermediate-emissions pathway that could lead to 2.3-4.6°C of global warming by 2300.

But that's if we actually manage to implement climate policies that are currently on the table. If the world fails to reduce greenhouse gas emissions, we have to consider the consequences of higher levels of global warming. Geologic records show that between 16.9-14.7 million years ago, during a prolonged warm period known as the Miocene Climatic Optimum, atmospheric carbon dioxide concentrations were 400-600 ppm, similar to those projected for the end of the twenty-first century under the "middle of the road" emissions scenario (SSP2-4.5). However, geologic records suggest that global temperatures were a staggering 5-10°C warmer than pre-industrial conditions. It is believed that Arctic sea ice may have been non-existent at times, and the Antarctic ice sheet was much smaller or perhaps even absent. It's a mind-boggling thing to consider, as Greenland alone stores enough ice to raise global sea levels by around 7 meters, with Antarctica currently locking away a further 58 meters of sea level. During this warm period, global sea level is estimated to have been 10-60 meters higher.

Under the worst-case scenario of continued fossil fuel growth and no additional climate policy (SSP5-8.5), carbon dioxide concentrations exceed 1000 ppm by the end of this century. The IPCC report warns that we could see warming of 6.6–14.1°C by 2300, a level not experienced since the Early Eocene Climatic Optimum some 50 million years ago. For context, we are talking about the greenhouse world of the early age of mammals. Independent geologic records suggest that this was a prolonged "hothouse" period when carbon dioxide concentrations were 1150–2500 ppm, similar to levels projected under the SSP5-8.5 scenario by 2100. The evidence suggests temperatures were 10–18°C warmer than pre-industrial conditions, with tropical forests growing as far south as 70°S in Antarctica. The poles were essentially ice-free, with global sea level estimated to be between 70 and 76 meters higher than present. It's a radically different world that would completely transform our thriving blue planet into a hothouse hell without humans.

*　　*　　*

While the scenarios described in this chapter read like something from a science fiction novel, unfortunately this material is the actual science contained in the latest IPCC report. The most important message here is that the risk of abrupt climate change increases with higher levels of warming. Right now, things are still in our hands, but the longer we delay, we run the risk of crossing critical tipping points that could see our world radically transformed in just a handful of decades. We are already seeing an escalation of extreme conditions like the rapid melting of the Arctic, the die-off of the Amazon rainforest and mass coral bleaching across the tropics – all clear signs that our world is rapidly changing. The truth is, we are likely to experience more unwelcome surprises as we know that our climate models are missing some important, non-linear processes like permafrost melt, wildfires and ice sheet disintegration that amplify warming, so these things are likely to catch us off guard. By looking over the Earth's geologic history, we can see that our current trajectory is leading us straight into dangerous levels of warming that will only worsen if we don't stabilize global greenhouse gas emissions immediately.

So far, we've looked at the physical changes expected with our warming planet, the nuts and bolts of what drives our climate and why it is changing. But the reality is that most people don't understand the world through temperature projections or disrupted ocean circulation patterns like scientists do. People want to know how our lives will change; they want to understand the threats to the landscapes we love, our culture, our homes. The intergenerational damage. To truly understand what is at stake, we need to move out of the realm of the head, and into the wisdom of our heart.

PART 2

The Heart

5

Where the battle has been lost

THE NATURAL WORLD IS WHERE we go to reflect, play and connect, like an animal feeding from its mother – primal, urgent, vital. For
me, there are few things I love more than an ancient forest. Their beauty
feels timeless, eternal. When you pass through these places, your
pores fill with a calm that recalibrates your senses, restores your sense
of peace. Their presence puts you in touch with the vast continuum of
time – a direct experience of our evolutionary history – reminding you
that your mortality is just a blip on the radar. Some individual trees can
live more than 3000 years, longer than the oldest pyramids of Egypt
have been standing. The oldest tree alive is called Methuselah, a gnarly
bristlecone pine (*Pinus longaeva*) in the White Mountains of eastern
California, named after the biblical figure believed to have been the
oldest person who ever lived. It's estimated to be around 4853 years
old, close to seventy times the average human lifespan.

The most incredible trees I have ever encountered are the kauri of
New Zealand (Aotearoa). *Agathis australis* is a primitive conifer that
belongs to the very ancient *Araucariaceae* family of trees that were pro-
lific before the southern super continent Gondwana broke up during
the Jurassic period more than 180 million years ago. These native pines
are among the oldest trees in the world, with some reaching ages of
more than 2000 years. They are the Southern Hemisphere's answer to

the giant sequoias (*Sequoiadendron giganteum*) of California, the largest trees in the world. The kauri is the biological cousin of Australia's celebrated Wollemi pine, an exceptionally rare "living fossil" discovered in 1994 in the Blue Mountains World Heritage Area west of Sydney. While only a small stand of Wollemi pines survives deep in an isolated canyon, its location a heavily guarded secret, northern New Zealand still contains tracts of magnificent kauri forest that you can visit with ease.

During my PhD research days, I spent two years based at the Tree Ring Laboratory at the University of Auckland, learning the painstaking craft of using patterns in kauri growth rings to reconstruct past El Niño events. Our team worked in remote groves of ancient forest on moss-drenched ridgetops and rugged slopes – exactly what you'd expect if you wandered into Tolkien's Middle Earth from *The Lord of the Rings*. We were there sampling trees to extend the Southern Hemisphere's climate record nearly 4000 years continuously back in time. One of the most amazing places I visited was the magical Waipoua Forest in Northland, home to two of the largest living trees left in New Zealand: Tāne Mahuta (Lord of the Forest) and Te Matua Ngahere (Father of the Forest). Tāne Mahuta is 51 meters tall and a staggering 4.4 meters wide. Although the tree's age is unknown, it is estimated to be around 2500 years old. Te Matua Ngahere is estimated to be over 2000 years old and is 5.3 meters in diameter – that's a tree about as wide as the length of a twelve-seater minibus. A true marvel of nature.

Today, forests cover only around a third of the global land surface, with more than half found in Brazil, Canada, China, Russia and the United States. In the past, the future of these extraordinary places seemed certain, but now remnant forests all over the world are facing increasing drought, wildfire, disease and human encroachment that threaten their existence. In recent years, many regions including the Mediterranean, western North America, southeastern Australia and central South America have experienced an increase in the severity of wildfires as the planet continues to warm. In September 2021, the US state of California experienced record-breaking fires that reached the Sequoia and Kings Canyon National Parks containing some of the oldest and largest trees left in the world. Firefighters draped giant sequoias in flame-retardant foil to protect the ancient trees from being engulfed by the wildfires that devastated the region. The world's largest tree, General Sherman, a towering 84-meter sequoia thought to be around 2300–2700 years old, made global headlines as firefighters rushed to guard the most iconic trees left in perhaps the most highly protected forests of North America. It's hard to believe it's come to this – wrapping individual trees in protective blankets to prevent them from being destroyed forever.

It reminded me of Australia's Black Summer bushfires of 2019–2020, when the world's last stand of prehistoric Wollemi pines was saved from being incinerated by the Gospers Mountain megafire that ignited in remote bushland northwest of Sydney. The unprecedented rescue mission involved water-bombing aircraft and large aerial drops of fire retardant. Helicopters winched in specialist firefighters to set up an irrigation system to wet the forest floor in an attempt to slow the advance of fire and dampen ember attacks. The crew worked in oppressive heat as horrendous conditions closed in on them. When the blaze finally tore through, we saw miraculous images of a tiny canyon, glowing impossibly green, surrounded on all sides by blackened wasteland. It was the only sign of life in a landscape completely decimated by fire; a feat only made possible because of active human protection. The symbolism made me weep.

Although the heroic effort prevented the significant loss of large, mature Wollemi pines, the canopy was severely scorched, badly burning many younger trees. It is still unknown whether they will resprout and fully recover following the catastrophic fires. The disaster prompted experts to call for permanent fire protection so they are not lost to humanity forever. Although I feel immense gratitude for conservationists who do everything humanly possible to save individual trees and iconic species, deep down I know that these heroic deeds will never be enough to save entire ecosystems. Unless we stabilize the Earth's climate, my fear is that climate change will outrun evolution in the end. We will be no match for the ferocity of the human-fueled disasters the future will bring.

*   *   *

The Earth's incredible biodiversity is the feature that distinguishes us from other planets. Our tangled jungles, teeming coral reefs, and windswept deserts contain a staggering array of life that scientists are still grappling to describe. Although measures of biodiversity vary considerably, taxonomic records indicate that there are around 8.7 million known species of plants and animals, which only represents a fraction of the total number of species that actually exist. Ecologists estimate that 86 percent of creatures found on land and 91 percent of the species that live in the ocean are yet to be described. It's an astonishing thing to consider – we have only just begun cataloguing the true extent of life on our planet. We know more about the moon than we do about our oceans.

Despite our incomplete knowledge, living organisms are recognized as the Earth's life-support system, and are humankind's shared evolutionary heritage. The concept of biodiversity takes in the array of plants, animals and microorganisms that form populations which, in turn, function collectively as ecosystems. These ecological units work together to continually cycle energy and materials through living organisms that grow, reproduce and decay. This cycling has evolved in response to a range of ecological processes of competition and predation over millennia, alongside other disturbances like disease and fire that can alter the trajectory of natural selection. Together, these processes provide what is termed "ecosystem services," delivering humanity clean water and air, natural carbon storage, and productive topsoil.

Aside from their utilitarian function, the natural world also has intrinsic value of its own. According to the Intergovernmental Science-Policy Platform on Biodiversity and Ecosystem Services (IPBES) – the ecological equivalent of the IPCC – many knowledge systems use the concept of "Mother Earth" to describe humanity's relationship with life. In this world view, a benevolent mother endows humans with nature's gifts that, when carefully protected, allow us to live in harmony and balance with the natural world. This concept of custodianship is intuitively understood by the world's Indigenous peoples, and over the past decade or so, has started to become enshrined in law in at least twenty countries around the world, marking the evolution towards recognizing the inherent rights of nature to exist, thrive and endure. For example, the Ecuadorian and Bolivian constitutions now recognize the legal rights of nature, known as Pachamama (Mother Earth). After a national referendum in 2008, Ecuador changed its constitution to reflect the right of nature to "exist, persist, maintain and regenerate its vital cycles."

Inspired by these developments in South America, in 2009 the United Nations adopted the *Resolution on Harmony with Nature* to acknowledge that the depletion of the world's natural resources and rapid and widespread environmental degradation are the result of unsustainable consumption and production practices that have damaged the Earth, along with the health and wellbeing of humanity. The resolution stated that the UN General Assembly was "convinced that humanity can and should live in harmony with nature." The principles of *Harmony with Nature* called for a move towards sustainable development that recognizes the rights of the natural world, and our shared existence on this planet. In 2014, New Zealand went one step further and granted legal personhood to the protected area of Te Urewera, the largest expanse of native forest left on the North Island, becoming the first country in the world to give nature the same rights as a human being. In Australia, the Yarra River in Melbourne was legally recognized as a living entity deserving protection in 2017. In the *Yarra River Protection (Wilip-gin Birrarung murron) Act*, the local Wurundjeri people stated: "The Birrarung is alive, has a heart, a spirit and is part of our Dreaming... since our beginning it has been known that we have an obligation to keep the Birrarung alive and healthy for all generations to come."

In contrast to this idea of Earth stewardship, many capitalist cultures see nature as a collection of raw materials that are there to be exploited to meet endless human demand. Some people use the Judaeo-Christian scriptures to justify the dominance of human beings over nature, laid out in chapter 1 of the Book of Genesis:

> Let us make humankind in our image, according to our likeness; and let them have dominion over the fish of the sea, and over the birds of the air, and over the cattle, and over all the wild animals of the earth, and over every creeping thing that creeps upon the earth... and God said to them "be fruitful and multiply... fill the earth and subdue it..."

As Australian social scientist John Wiseman outlines in his inspirational book *Hope and Courage in the Climate Crisis*, modern Christians are moving away from this destructive notion, citing Pope Francis's 2015 encyclical on climate change and ecology, *Laudato Si': On Care for Our Common Home*. In his deep contemplation of the ecological crisis, addressed to "every person living on this planet," the Pope called on the church and society at large to acknowledge the urgency of the environmental crisis:

> Praise be to you, my Lord, through our Sister, Mother Earth, who sustains and governs us, and who produces various fruit with coloured flowers and herbs. This sister now cries out to us because of the harm

we have inflicted on her by our irresponsible use and abuse of the goods with which God has endowed her. We have come to see ourselves as her lords and masters, entitled to plunder her at will. The violence present in our hearts, wounded by sin, is also reflected in the symptoms of sickness evident in the soil, in the water, in the air and in all forms of life. This is why the earth herself, burdened and laid waste, is among the most abandoned and maltreated of our poor; she "groans in travail" (Rom 8:22). We have forgotten that we ourselves are dust of the earth (Gen 2:7); our very bodies are made up of her elements, we breathe her air and we receive life and refreshment from her waters.

Pope Francis goes on to cite an alternative paradigm for humanity's relationship with nature found in the teachings of the thirteenth-century holy man Saint Francis of Assisi, who believed in the equality of all beings. In opposition to the exploitative interpretation of humanity's relationship with nature, Saint Francis spoke of the ecological stewardship and interdependence of humans and the natural world. In his moving reflection on the ecological crisis, Pope Francis writes of the modern-day relevance of centuries-old wisdom:

I believe that Saint Francis [of Assisi] is the example par excellence of care for the vulnerable and of an integral ecology lived out joyfully and authentically. He is the patron saint of all who study and work in the area of ecology ... He shows us just how inseparable the bond is between concern for nature, justice for the poor, commitment to society, and interior peace ... Francis helps us to see that an integral ecology calls for openness to categories which transcend the language of mathematics and biology, and take us to the heart of what it is to be human ... if we feel intimately united with all that exists, then sobriety and care will well up spontaneously.

As a scientist with no religious leanings, I was moved to learn that an alternative vision for our relationship with the Earth has long existed within spiritual philosophy. It was also good to know that we have a patron saint looking out for us, as we need all the help we can get. This gentler view has just been overshadowed by more aggressive, self-serving forces of capitalism that have dominated Western social and economic values for far too long. It gives me hope that if we recast our relationship with the natural world as one of custodianship and interdependence, one that speaks to a deeper, shared truth that transcends our divisions, we can restore our rightful place as caretakers of this planet. Ultimately, Pope Francis's message is a powerful call to action, one intended to unite across cultures, time and faiths:

> I urgently appeal, then, for a new dialogue about how we are shaping the future of our planet. We need a conversation which includes everyone, since the environmental challenge we are undergoing, and its human roots, concern and affect us all ... Regrettably, many efforts to seek concrete solutions to the environmental crisis have proved ineffective, not only because of powerful opposition but also because of a more general lack of interest. Obstructionist attitudes ... can range from denial of the problem to indifference, nonchalant resignation or blind confidence in technical solutions. We require a new and universal solidarity. As the bishops of Southern Africa have stated: "Everyone's talents and involvement are needed to redress the damage caused by human abuse of God's creation." All of us can cooperate as instruments of God for the care of creation, each according to his or her own culture, experience, involvements and talents.

Amen.

*   *   *

To appreciate the reality of living on a warming planet, we need to understand the associated ecological crisis we are facing. Changes in the places we love are how we directly experience the natural world in our daily lives. Our private communion with nature takes place on walks along riverbanks, pilgrimages up high mountain tops, and immersion in revitalizing seas. It also takes place in our relationships with our local parks and gardens, our pets, and what's on our plates. When we start to understand how changes in the web of life are impacting our daily lives, we make contact with our instinctive response to protect our earthly home, and with it, start to restore our inner peace.

Before humans cleared vast tracts of nature to make way for agriculture, urban settlements and infrastructure, intact ecosystems thrived, supporting a diverse array of plants and animals across a diversity of habitats. As the size of the human population increased to 7.9 billion, so too has our demand for resources, placing immense pressure on the natural world. Extraction of natural resources has more than tripled since 1980. According to the IPCC's Working Group 3's *Mitigation of Climate Change* report, increases in the global demand for products and services is the strongest driver of greenhouse gas emissions, accounting for up to two thirds of total emissions when all parts of the process are considered. Our insatiable needs have led to the destruction of natural habitats the world over, causing the extinction of species and the overall decline in ecosystem function and diversity. As of 2020, less than a quarter of the global land surface still functions in a near-natural state with its biodiversity still largely intact. Put another way, we have now disturbed and degraded more than three quarters of our planet.

Conservation efforts by dedicated humans have resulted in the protection of globally significant areas of ecological importance. A "biodiversity hotspot" is an area that is exceptionally rich in species, is ecologically distinct, and contains a high number of unique species found there and nowhere else, a characteristic referred to as biological endemism. While some of these regions are protected in World Heritage Areas, national parks or internationally recognized conservation reserves, these hotspots only make up around 16 percent of global land area. This tiny fraction of our planet houses around half of the world's plant and one third of vertebrate species, despite the fact that around 85 percent of the original extent of these areas has been destroyed. Disturbingly, between 1992 and 2015, the world's most biologically rich areas lost nearly 148 million hectares, an area more than three times the size of Sweden, primarily to agriculture and urban sprawl. The top driver of continuing deforestation is grazing land used for commercial cattle. Brazil alone devotes an area seven times the size of the United Kingdom to beef production; much of this was once part of the largest rainforest left in the world.

According to the IPCC's Working Group 2's *Impacts, Adaptation and Vulnerability* report, the risk of extinction for all native animal and plant species in land and marine biodiversity hotspots increases in all future climate change scenarios, with up to a quarter of species at very high risk of extinction in these precious places above 1.5°C of warming. The threat of extinction is even higher for endemic species in these regions, rising to 100 percent of species on islands, 84 percent on mountains, 54 percent in the ocean and 12 percent on land. These stark figures indicate that biodiversity hotspots in the current climate will not be refuges from the effects of global warming in the future. Some species will have nowhere to go, as they can't retreat to find cooler conditions further up mountain slopes or disperse into alternative habitats that now neighbor human settlements and industrial-scale farms. The IPCC warns that even temporarily exceeding 1.5°C of global warming will result in severe impacts, some of which will be irreversible in vulnerable mountain, polar and coastal ecosystems.

The IPBES's 2019 *Global Assessment Report on Biodiversity and Ecosystem Services* describes how threats to natural systems across the globe that have already been significantly altered by human disturbance are now being compounded by climate change. Seventy-seven percent of the Earth's land surface has been significantly modified by humans, with over 15 billion trees cut down every year. The global number of trees has nearly halved since the start of human civilization. The world's rainforests have also been slashed by half, particularly in the tropical nations of Indonesia, Brazil and the Democratic Republic of Congo. Seventy percent of the birds on Earth are now domesticated; the vast majority are chickens. Globally, humans consume 50 billion of them each year; many of these birds are fed with soy-based feed grown on deforested land. Over the past decade, land degradation contributed to around a quarter of the world's greenhouse gas emissions, with over 50 percent derived from deforestation.

The situation on the coast is not much better. Only 13 percent of the ocean remains free from our impacts, with humans removing 80 million tonnes of seafood from the oceans each year. Around 90 percent of the world's largest fish like giant blue marlin, southern bluefin tuna and swordfish have been removed by industrialized fishing. Plastic rubbish has increased ten-fold since 1980, and is now found throughout the entire ocean, from surface waters to deep ocean trenches, forming giant garbage patches in places like the North Pacific. Eighty-seven percent of the world's natural wetlands – biologically diverse systems critical for maintaining water quality, reducing the impacts of storm surge and flooding, and naturally storing carbon – have been destroyed, with a third lost since 1970, mainly to support commercial agriculture and fisheries. Offshore, approximately half of the world's living coral cover has been lost since the 1950s. Many coral reefs – the "rainforests of the sea," home to a quarter of all marine species – are now in terminal decline. Fertilizers entering coastal ecosystems have resulted in more than 400 ocean "dead zones," with a total area larger than the United Kingdom.

And if that isn't enough, wild animal populations have halved since the 1950s. Even more startling is the fact that 96 percent of the mass of all the mammals on Earth is made up of humans and the animals we eat. Our own mass accounts for one third of the total, while domesticated animals such as cows, pigs and sheep make up 60 percent. The remaining 4 percent is made up of wildlife like elephants, whales and monkeys – it's shocking to realize that less than 5 percent of all the mammals left on our planet are truly wild.

A recent study by Andrew Plumptre from the University of Cambridge estimates that this scale of destruction now means that only 3 percent of the Earth's land ecosystems can now be considered ecologically intact. As the great British naturalist David Attenborough explains:

> We have forgotten that once there were temperate forests that would take days to traverse, herds of bison that would take four hours to pass, and flocks of birds so vast and dense that they darkened the skies. Those things were normal only a few lifetimes ago. Not anymore. We have become accustomed to an impoverished planet. We have replaced the wild with the tame. We regard the Earth as our planet, run by humankind for humankind. There is little left for the rest of the living world. The truly wild world – that non-human world – has gone. We have overrun the Earth.

It's an appalling reality to come to grips with. Nature is being destroyed at a shameful rate – what's left is finite and needs our protection. The Secretary-General of the United Nations, António Guterres, perfectly sums up what so many of us feel:

> Humanity is waging war on nature. This is senseless and suicidal. The consequences of our recklessness are already apparent in human suffering, towering economic losses and the accelerating erosion of life on Earth.

The cumulative pressure of our damage is causing natural systems to buckle and collapse: it is happening right before our eyes, yet most of us fail to see it. Ecologists recognize that there are tipping points in the complex systems of nature. Once a critical threshold is reached, it triggers sudden changes that eventually stabilize in a new, altered state. The environmental stability we have taken for granted throughout the entire course of human history is now faltering. The ecosystem services that nature has provided for free since the dawn of time are now breaking down. Given that around 56 percent of carbon dioxide emissions from human activities are physically and biologically absorbed by the land and oceans, maintaining the environmental integrity of the natural world is critically important for controlling the level of climate change that we experience. It is also vital for maintaining a functional biosphere; that is, life on Earth. The IPCC report warns that under scenarios with increasing greenhouse gas emissions, the ocean and land become less effective at slowing down the accumulation of carbon dioxide in the atmosphere, further warming our planet and threatening ecosystem functioning.

Right now, the very foundation of life on our planet is increasingly at risk because of relentless human pressures that are being made worse by global warming. While species have gone extinct throughout time, human activities, including habitat loss, commercial exploitation and climate change, are accelerating this process. A 2020 study led by Cristian Román-Palacios from the University of Arizona calculated that up to one third of all assessed plant and animal species could be extinct by 2070 if climate change continues at current rates. Under a high-emissions scenario, more than half of the 538 analyzed species could be lost, even when accounting for shifts in dispersal and niche characteristics. The latest IPCC estimates – based on a global assessment of tens of thousands of species – suggest that up to 18 percent of land-based species could become extinct at 2°C of warming, wiping out as much as a third of all species if 3°C is reached. Close to 70 percent of species in biodiversity hotspots were found to be at either "high" or "very high" risk of extinction due to climate change. Threats are especially severe for endemic species restricted to islands and mountains, where over 80 percent of species could perish.

With 2°C of warming, which the IPCC says could be reached as early as the 2040s, polar species like penguins, seals and bears will be threatened, while tropical coral reefs and mangroves will degrade substantially. Ninety-nine percent of the world's tropical coral reefs will die off, condemning marine wonders like Australia's Great Barrier Reef and the Caribbean's Mesoamerican Reef to the history books within our lifetime. In South America, only 1 percent of the Pantanal, the largest tropical wetland in the world, remains viable in an area that currently boasts the highest concentration of wildlife on the continent, including jaguars, toucans and giant otters. In the wet forests of the central Congo in equatorial Africa, a region that contains the world's second largest tropical rainforest housing a stunning array of primates including gorillas and chimpanzees, only around 21 percent of the area will remain a suitable refuge at 2°C of global warming. In the biodiversity hotspot of southwestern Australia, which contains about 8000 plant species – more than one third of all Australian plant species – only 11 percent of the area is projected to remain a sanctuary for the region's abundant life. That area shrinks to just 5 percent above 3°C.

With 4°C of warming, which is projected to be reached around the 2080s under high-emissions scenarios, mass mortality and extinctions will irreversibly alter life on the planet, greatly impacting the survival of biodiversity hotspots. In the Sierra Nevada Coniferous Forests of western North America, a mountainous region that contains some of the most significant groves of giant sequoias in the world, less than 20 percent of the area is projected to remain climatically suitable to support its exceptional life forms. In the Eastern Australia Temperate Forests hotspot, where more than 8250 species of plants – including the critically endangered Wollemi pine – are found, the percentage shrinks to a mere 13 percent. Under such high levels of warming, mangrove areas around the globe will be submerged, disappearing completely in places like New Guinea, Indonesia and the Sundarbans in the Bay of Bengal. In the Great Sandy-Tanami desert region of Central Australia, encompassing areas including the sacred Aboriginal site of Uluru–Kata Tjuta (Ayers Rock), the percentage of land that is projected to support current biodiversity shrinks to zero beyond 2.5°C of warming. It's impossible to fathom what this means for the local Anangu people, members of the oldest continuous culture in the world, who have inhabited this region of Australia for more than 60,000 years.

The United Nations Environment Programme's 2021 report *Making Peace with Nature* highlights that the proportion of the global land surface where ecosystems will shift from one biome to another as a result of climate change is projected to be around 8 percent for 1.5°C warming, 13 percent at 2°C warming, 28 percent with 3°C warming, and 35 percent at 4°C warming. That is, vast tracts of land will transition from their current conditions into different states to adapt to a new climate. Climate change increases the threat of extinction across the planet, with 20–30 percent of plant and animal species at increased risk under 2°C warming, with progressively more species threatened at higher levels of warming. The global fraction of insects, birds, mammals and plants that are projected to lose more than half of their current range by the end of the century is 4–8 percent for 1.5°C warming, 8–18 percent for 2°C, 26–49 percent for 3.2°C and 44–67 percent with 4.5°C. Such extensive losses will significantly undermine the provision of ecosystem services at a global scale, impacting processes like pollination that are needed to maintain land-based ecosystems, including those required for food production.

The IPCC's latest global assessment of thousands of species confirms that between half and two thirds of all species spread across terrestrial, freshwater and marine systems have already shifted their geographic ranges in response to the 1.2°C of global warming we've experienced so far. The report also highlights a study of 976 plants and animals that found that 47 percent had already suffered local extinctions as a result of climate change, warning: "These extinctions will almost certainly increase as global climate continues to warm in the coming decades." The IPCC also concludes with high confidence that "new analyzes demonstrate that prior reports underestimated impacts due to complex biological responses to climate change" – an admission that scientists are struggling to document rapid changes already being experienced in the natural world.

It is clear that climate change is altering the fundamental viability of life on our planet right now. We are already witnessing nature disappear at staggering rates, and that's with just 1.2°C of warming we have experienced so far. If the net-zero pledges announced during the UN Conference of Parties (COP) 26th meeting in Glasgow (COP26) in November 2021 are fully implemented through strict energy and land use policies – in a best-case scenario – the world could warm 1.4–2.8°C by the end of the century, with a central estimate of around 2°C. The problem is that some countries have near-term 2030 pledges that are far higher than business as usual projections and most are not legally binding, meaning that our future is still far from guaranteed. Even if the governments of the world make good on their promises, up to a third of life on Earth could become extinct, so it's clear that what we are doing is still not enough. This level of warming would spell the end of vast tracts of our biodiversity hotspots: tropical coral reefs, abundant wetlands and lush rainforests will be obliterated. It's hard to comprehend the level of destruction that will be unleashed as nature struggles to re-establish its equilibrium as the planet continues to warm.

I mourn all the unique animals, plants and landscapes that will be forever altered by our rapidly changing climate. It's very hard to come to terms with the fact that the Earth as we now know it may soon no longer exist. As British journalist George Monbiot writes in *This Can't Be Happening*:

> I have lived long enough to witness the vanishing of wild animals, butterflies, mayflies, songbirds and fish that I once feared my grand-children would experience: it has all happened faster than even the pessimists predicted. Walking in the countryside or snorkelling in the sea is now as painful to me as an art lover would find her visits to a gallery, if on every occasion another Old Master had been cut from its frame.

I grieve for the generations of children who will only ever experience the world's natural wonders through museum exhibitions or David Attenborough documentaries. In the future, his films will no longer be a celebration of the stunning life on our planet, but archival footage of a lost world. To give nature a fighting chance, we must do everything we can to slow the rate of warming to give evolution time to adapt. But as recent years have shown us, extreme weather events like record-breaking heatwaves or catastrophic wildfires can strike suddenly, altering our ecosystems faster than scientists ever imagined.

*   *   *

Many people outside of Australia have only ever heard of our country in the context of horrific bushfires, biblical floods and epic droughts. If they are adventurous, they may have visited our shores to soak in our natural beauty; stunning landscapes, unique wildlife, our glittering, waterside cities. But dig a little deeper and you'll find that we've earned ourselves an international reputation for being a terrifyingly fierce country – a wild card. It's a fair assessment; Australia is a land defined by dramatic extremes that often make global headlines.

Although Australia is a developed country with a stable govern-ment, it shares many of the environmental vulnerabilities experienced by the developing world. The nation is highly exposed to the vagaries of the weather; its key industries of agriculture and tourism, along with its infrastructure and cities, are incredibly vulnerable to rainfall and tem-perature extremes. Yet Australia is one of the few remaining developed nations that refuses to give up fossil fuels or to implement scientifically credible climate policy, even though the rest of the world has galloped on. The Australian government relentlessly tries to defend the unjusti-fiable position of jeopardizing our planetary stability to keep the fossil fuel industry alive for as long as possible. But more on that later.

Before the coronavirus pandemic swept the world, the scientific community was reeling from the most catastrophic bushfire season in Australian history. We watched on in horror as the fires savaged our country, releasing more carbon dioxide in a single bushfire season than the country emits in an entire year. Although eastern Australia's forests are among the most fire-prone in the world, typically only 2 percent burns during extreme fire seasons. In 2019–2020, a staggering 21 percent of Australia's temperate forests burnt in a single bushfire season, setting a new global record for the sheer enormity of the blazes. It's a formidable feat, particularly for a country with such a ferocious history of disasters.

Australia's Black Summer fires had unprecedented impacts over 24 million hectares – an area equivalent to the size of the entire United Kingdom. The Gospers Mountain megafire near Sydney alone burned more than half a million hectares, making it the largest forest fire recorded in Australian history. Together, the fires destroyed over 3000 houses and killed thirty-three people. Over 3 billion animals were incinerated or displaced by the blazes. A terrifying amount of Austra-lia's World Heritage Areas were burnt – at least 80 percent of the Blue Mountains protected area and 53 percent of the ancient Gondwana rainforests network. These are the "last of the last" of such precious places. Areas that have clung on since the age of dinosaurs, forced to contend with the processes of evolution playing out in fast forward.

Instead of adapting gradually over thousands or millions of years, ecosystems were radically transformed in the space of a single summer – not even a nanosecond in geologic time. Some of our most precious ecosystems may never recover. Gutted landscapes will struggle on, trying to regain some semblance of an equilibrium, but the truth is the destruction we have unleashed will reverberate throughout the ages. The legacy of these fires will be intergenerational. Ecologists openly wept as they surveyed the monstrous extent of the destruction in the aftermath of the event.

The climate disruption caused by the Black Summer wildfires of 2019–2020 ruptured every Australian's sense of personal safety, undermining our assumption that life will continue on as it always has. The truth is, climate change is happening so shockingly fast in our country – scientists are struggling to reconcile the gulf between the inconceivable impacts of recent record-smashing extremes, and the model projections that look ridiculously underestimated as each season passes. We are yet to come to terms with what future life in our country will look like, but that horrendous summer gives us a pretty graphic preview of what we can expect.

On 3 January 2020, the Australian army evacuated over 3000 people from the bushfire-ravaged town of Mallacoota in eastern Victoria, in the southeast corner of the country. It was the largest maritime evacuation in Australian history. Holidaymakers were forced to abandon their cars, complete with kids' bikes strapped to the roof racks, ice melting in cool chests. People hoping for a carefree break over the new year were instead in the unimaginable position of having to flee for their lives. Tens of thousands of people in coastal New South Wales and Victoria were stranded in fire-ravaged towns where the highways were closed, supermarkets ran out of food, and queues for petrol snaked down the streets. Those caught up in the ordeal described the scenes as apocalyptic.

Meanwhile, local residents faced the infinitely more serious situation of returning to find their homes completely incinerated. Cars melted, pets killed, beloved landscapes destroyed. A lifetime of memories razed to the ground. As Australia's climate continues to warm, the most intimate places of human safety – our very homes – are being threatened by an increasingly dangerous world.

It was surreal to witness military evacuations, usually reserved for developing regions of the world following natural disasters, deployed in twenty-first-century Australia. The sheer scale and severity overwhelmed our capacity as a nation to deal with the emergency. The problem was that it wasn't just one event impacting one region, but multiple disasters simultaneously unfolding in every state and territory around the country. If a wealthy nation like Australia is so unprepared, it's disturbing to think how the developing world will cope.

Yet when disaster struck, the Australian government refused to acknowledge that our climate is changing because human activity has so thoroughly altered the chemistry of our atmosphere and oceans that the entire planetary balance has been knocked off kilter. They failed to accept the fact that we are now experiencing an upheaval of life as we know it, not in some distant future, but right now, each and every day. As an Australian scientist involved in the IPCC process, I often felt like my country was a real-time poster child for the climate change impacts our group was so meticulously trying to document. Australia is living proof of how fast the world is changing.

The destruction caused by a rapidly warming planet is now painfully clear for the world to see. It feels like an invisible threshold has been breached, like an abrupt switch has been flicked, and we now find ourselves in new terrain, somewhere off the map. It has been heartbreaking to witness the destruction of many of our UNESCO-listed ecosystems, areas that are recognized as having such "outstanding universal value to humanity" that they are inscribed on the World Heritage List to be protected for future generations. In the handful of years since 2016, the Greater Blue Mountains Area, the Gondwana Rainforest Network of Australia and the Great Barrier Reef – three of our most iconic World Heritage sites – have been utterly devasted by climate change.

In late March 2020, just before Australia's national coronavirus lockdown took effect, Professor Terry Hughes, one of the world's leading experts on coral reefs and our foremost authority on the Great Barrier Reef, rushed to conduct an aerial survey of the third mass bleaching event to strike the reef since 2016. It was the first time that severe bleaching impacted virtually the entire reef, including large parts of the southern reef that had been spared during bleaching events in 2016 and 2017. It was hard to hide from the reality that the entire system is in an advanced state of ecological collapse. In desperation, Terry took to Twitter, sharing his experience of surveying the carnage: "It's been a shitty, exhausting day on the #GreatBarrierReef. I feel like an art lover wandering through the Louvre . . . as it burns to the ground." By the end of his fieldwork he was a broken man: "I'm not sure I have the fortitude to do this again." His results now reveal that just 1.7 percent of the world's largest coral reef has escaped mass bleaching since 1998.

For many scientists, it has been shattering to realize that we are likely to be the last generation that experiences the world as we know it today. I never thought I'd live to see the loss of some of the planet's greatest ecosystems; our shared natural inheritance, passed on from one generation to the next, since time immemorial, is no longer guaranteed. In the world's wild places, it's easy to reconnect with a sense of timelessness, to feel what our ancestors once felt. To remember that these landscapes have borne witness to an eternity of inner and outer world struggles, silently offering us the wisdom and restoration that come from being still and listening. We must re-establish our connection with the natural world, restore our care for the Earth, and with it our humanity.

*   *   *

When the world seems like it is disintegrating all around you, sometimes the only sane thing to do is to turn inward and try to regain your bearings, one moment at a time. By focusing on what you can control, however minuscule and inadequate that might feel: the depth of your breath, your exposure to distressing information, your connection with people that keep you anchored in rough seas. Although it feels impossible to continue on as normal when the world has become surreally deranged, somehow we must endure and try to find meaning in our private places. To tap into unknowable forces that somehow sustain us, that gently remind us that life must go on.

As the Black Summer wildfires consumed the subtropical rainforest inland from where I live, we packed uneasily for our honeymoon in Japan. My life had been on overdrive, so we decided to take the holiday first before getting married on a secluded beach close to home. Just us, the ocean, and a surfer out for a dawn paddle as our witness.

Leaving Australia in the middle of the bushfire crisis felt like abandoning my post. Messages from my media contacts went unanswered. By early December 2019, the situation had deteriorated rapidly; people wanted me to help them make sense of what was going on. But to do that, I had to face the enormity of the disaster myself, and I just wasn't ready. After a text message from a producer at the BBC World Service, I decided to turn my phone off and try to avoid reading the news while I was away. Thinking about climate change permeates my daily life – but surely I could take a break, just for a little while. I needed to recharge before returning to face another round of IPCC edits to meet our 12 January deadline.

After a few days in dazzling Tokyo, we traveled to the town of Tanabe, one of the starting points for the Kumano Kodo: an ancient pilgrimage trail through the remote mountains of the Kii Peninsula, south of Osaka. The region, considered the spiritual heartland of Japan, contains many Shintō shrines dedicated to the worship of nature. Natural objects like trees, mountains, rocks and rivers are considered deities. People also believe that the spirits of the dead congregate in the peaks, descending to commune with the living as they pass through the landscape. It's a sacred place of healing, where pilgrims go to be spiritually and physically purified. Walking through the mountains eventually undid me; the place had such a haunting quality that brought forth deep, uncontained sorrow. It was impossible to stop the river of tears moving through me.

Hiking through the Japanese forest made me appreciate the immense tragedy of the destruction back home. When you walk through the Australian bush or through one of our ancient rainforests, it's a transcendent experience. There is so much life all around you, even in the harshest of places. Our landscapes are so alive; in summer, the thrum of life can be deafening. After thousands of years of human disturbance, in Japan, you are lucky to see a handful of birds in the forest. The only wildlife we saw there was roadkill. The incredible diversity and abundance of Australia's natural landscapes are globally unique. It broke my heart to think of all the birds, koalas, possums – all of our extraordinary creatures – fleeing for their lives. The Blue Mountains World Heritage Area, where I'd addressed a packed hall just two weeks before, was now on fire. Iconic vistas were now glowing red. In the end, 80 percent went up in smoke. Writing in my journal that night, I tried to make sense of the feelings that overwhelmed me on the trail that day:

> It's so profoundly sad to think of how much has been lost, what is disappearing before our eyes. I never thought I'd live to see the day when the horror of all of this would begin to unfold. And yet, here we are,

on the threshold of a rapidly destabilizing planet. Balancing the grief I feel in my outer and inner worlds is my greatest challenge. To somehow still maintain some sort of joy and hope in this human existence. To not be engulfed by despair, depression and the crushing sense of responsibility I feel to do all that I can to bring awareness to this great unraveling.

*   *   *

After our week communing with the nature spirits of the Kumano Kodo, we headed to the ancient city of Kyoto and checked into a contemporary art gallery hotel in the quiet backblocks near the Komo River. Each room is part of the permanent art collection, housing exquisite works. We were placed in a room featuring local artist Showko, a ceramicist skilled in kintsugi; the Japanese artform of repairing broken pottery with lacquer laced with powdered gold. Instead of trying to conceal breaks, kintsugi lovingly mends each jagged seam, transforming the original object into a profound symbol of healing. It offers a meditation on what it means to come to terms with brokenness, to find joy and new meaning in the aftermath of loss. Scars are a celebrated symbol of resilience, of life holding on.

It was comforting to find myself contemplating kintsugi at a time when I felt so broken-hearted, so bereft. In recent months, I'd found myself coming apart easily. I'd been processing difficult things in my inner world, with my IPCC work constantly putting me in touch with unbearable realities. Depression was taking hold again, but my frenetic pace often means I don't have much time to stop and process things. Perhaps staying busy helps me cope with the distress that I feel. It's probably a way of convincing myself that I'm doing everything that I can. A relentless diary is a good distraction from the world disintegrating around you, but it won't save you when complete sorrow eventually soaks through.

My encounter with kintsugi was an epiphany – I realized that perhaps the wisest thing we can do is accept that life is full of immeasurable loss and pain. When you eventually come to terms with its heartbreak, it's easier to recognize the beauty, something worth salvaging from the wreckage. There is always someone willing to walk through the rubble, seeking out signs of life, something worth saving. There is power in knowing that we have a choice about how we respond to the distressing realities we now face, that we can choose to heal our world. Because ultimately, we all have the same choice to make – to step up and leave a legacy, or to ignore the call of history. It really is as simple as that. You can choose to be a person who restores faith in the goodness of humanity, to be someone who chooses to care, to repair, to hold on, even when all feels lost.

6

Sea of humanity

AS THE NATURAL WORLD HAS been cleared to make way for humans, our blue planet has become a concrete jungle. Thousands of cities now cover the surface of the Earth, housing around 4.2 billion people, more than half of the world's population. Cities account for two thirds of global energy consumption and around 70 percent of greenhouse gas emissions. They also consume around 70 percent of the world's food. Around a quarter of all cities have more than 5 million residents, roughly the size of Melbourne, Barcelona or Singapore. Once a city is built, its physical form is locked in for generations, leading to urban sprawl that further encroaches on natural landscapes as housing demand continues to grow. When cities swell to over 10 million people they are classified as "megacities"; the most populous city on the planet is currently Tokyo, coming in at a whopping 38 million. If you have ever been there and navigated the swarming train stations of Shinjuku or Shibuya, you'll know what it feels like to be part of a vast sea of humanity.

Megacities currently contain around 13 percent of the total human population, with most of these areas found in Africa, Asia and Latin America. This includes the cities of Tokyo, Delhi, Shanghai, São Paulo, Mexico City and Cairo, which all have populations greater than 20 million. An additional 2.5 billion people are projected to live in urban areas by the middle of the century, mostly in countries like India, China and Nigeria. Many of these areas already face the challenges of poor air quality, congestion and water insecurity – and that's without considering the added pressure of climate change.

Today, around a quarter of the world's urban population lives in slum conditions, predominantly in developing nations. These impoverished areas are made up of makeshift structures that lack a formal supply of basic infrastructure and services like safe water, sanitation, waste collection, adequate space and security. They often contain low-income, marginalized and displaced people who live in dilapidated housing in hazardous areas like flood-prone riverbanks and swamps, or near industrial factories and rubbish dumps, usually on the outskirts of town. Despite efforts to improve living conditions in slums and prevent their formation, the number of people living in these areas continues to grow. This reflects increasing urbanization and population growth rates, which are outpacing the construction of affordable housing over much of the developing world.

It's also interesting to know that some of the slums that exist today are the legacy of colonial rule. For example, Europeans arrived in Kenya during the nineteenth century and created urban centers like Nairobi to serve their financial endeavors like tea and coffee plantations. Locals were regarded as a supply of cheap labor, so employers and government authorities built settlements to house workers, who often came from rural areas. Due to the time and money spent traveling between cities and home, workers' entire families eventually migrated to urban centers. As these low-income workers could not afford to buy houses, urban slums were formed.

Other slums, like Dharavi in Mumbai, India, were created because of segregation imposed by the colonialists. In 1887, the British colonial government exiled all tanneries and other noxious industries – along with local poor people from the peninsula – to the northern fringe of the city, providing no investment in infrastructure, sanitation, public services or housing. Rural migrants then began moving into the area to find work as servants in colonial offices and homes, or in foreign-owned tanneries and other polluting industries like textile factories. The poor built makeshift shanties within easy commute to work, laying the foundation for what is now the largest slum in India.

It is estimated that 70 percent of the world's population will live in cities by 2050, with many people facing life in informal settlements and slums across Africa and Asia. While most countries do not generate formal statistics on the number of people living in these areas, the United Nations estimates that over 1 billion people live in slums. When you include different types of informal settlements, experts believe this number could exceed 1.6 billion. The situation is particularly bad in sub-Saharan Africa, where around 60 percent of the urban population lives in impoverished areas of Kenya, Nigeria and South Africa. In the case of Kenya's largest slum, Kibera, in Nairobi, sanitation is very limited, with open sewer lines emptying effluent in front of people's houses. There are only 1000 public toilets to serve the entire slum population of around 700,000. As a result, public defecation areas are still common, increasing the risk of outbreaks of cholera and typhoid.

In South Asia, around 30 percent of the urban population is crammed into informal settlements and slums in India, Pakistan and Bangladesh. In the Dharavi slum, where the British movie *Slumdog Millionaire* was filmed, an area of just over 2 square kilometers is estimated to contain a population of around 1 million people. The UN has estimated that there is an average of one toilet for every 1500 people, resulting in severe public health problems. Mahim Creek is a local river that is widely used by residents as an open sewer, causing the spread of contagious diseases. The residents have a section where they wash their clothes in the same water people toilet in, an indignity that symbolizes the shameful way we still treat the poorest of the poor.

Even in this modern era, one third of people in Asia still do not have access to safe, sustainable water supplies, and half do not have access to adequate sanitation. The UN estimates that over 2.2 million people in developing countries die from preventable diseases associated with lack of access to safe drinking water, inadequate sanitation and poor hygiene. Around 1.8 million people die each year from diarrhea and other preventable diseases related to unclean water, with children under five most affected. It's hard for people living comfortable lives in developed countries to fathom the extreme squalor and misery faced by millions of people, day in and day out, right across the developing world. Millions are already living in unimaginably wretched conditions; climate change is going to make a bad situation even worse.

The number of people living in urban areas who are expected to be highly exposed to climate change impacts increases with higher levels of warming. As the planet continues to warm, future generations will be exposed to higher risk of water scarcity, flooding, heat stress, poverty and hunger, all of which are made worse in urban settings, with children identified as the most vulnerable. According to the IPCC, the projected extent of urban land exposed to floods and droughts is expected to nearly triple from 2000 levels by 2030. Currently, over half of the world's population – around 4 billion people – already face water security issues related to climate change–induced water scarcity, population demand and inadequate water management. Model projections suggest that an additional 350 million people living in urban areas will be exposed to water scarcity from severe droughts at 1.5°C of warming, increasing to 410 million at 2°C.

Water security issues are worst in South and East Asia, sub-Saharan Africa and the Middle East. Impacts are particularly severe in Asia, where a quarter of a billion people currently rely on glaciers for water. Under the intermediate-emissions scenario the world is currently tracking, glaciers are expected to halve in the high mountain regions of Asia and shrink by up to 70 percent in central Asia by the end of the century. This has major consequences in the Himalayan glacier-fed river basins of the Indus and Ganges, where millions of farmers depend on snow and glacier melt to grow crops to feed the crowded cities of India and Bangladesh. These are monumental figures to try to take in – one of the most densely populated regions on Earth will have its water supply at least halved within the lifetime of children alive today. What's worse is that the IPCC reports that many glaciers around the world, not just in Asia, are projected to lose most of their mass or disappear completely regardless of future emissions. So much heat has already accumulated in the climate system from past emissions that some of these changes are now irreversible, especially for many low-elevation and small glaciers in places like the Andes, the North American Rockies and New Zealand. It's confronting to realize that some of these areas will lose most of their total mass even with 1.5°C of warming, which is expected to be breached in the early 2030s.

Globally, between 800 million and 3 billion people are projected to experience chronic water scarcity due to drought associated with 2°C of warming, with this number increasing to around 4 billion with 4°C of warming. South America will face an increasing number of days with water scarcity and restricted water access, especially for those living in cities like Santiago, Lima, Quito and La Paz, as the Andean glaciers continue to melt. On average, glaciers in the tropical Andes have already declined 30 percent over the past fifty years (with the figure as high as 60 percent in the Southern Andes). Low-elevation glaciers in Peru are at risk of halving by 2050 even under low-emissions pathways. Under a worst-case scenario, only around 5 percent of glaciers on the highest peaks of the tropical Andes will remain snow-capped by the end of the century, taking vital water supplies away from local communities.

In the arid nations of the Middle East, water shortages exacerbated by hotter temperatures are expected to have severe impacts on agriculture that will undermine food security in places like Saudi Arabia and Yemen. Climate change has already led to the forced phase out of Saudi Arabia's wheat production in 2016 due to a lack of water, leaving the nation reliant on imports. As water availability begins to dry up, we will continue to see conflicts over water and instability in food supplies that will lead to the migration of the rural poor into overcrowded urban areas in many parts of the world. But when they get there, they will find themselves in congested, polluted cities where they have traded one set of problems for another.

*   *   *

If you ever wander through a city on a warm summer's night, you'll notice that the surfaces around you radiate heat. If you stop and lean against a stone wall, you can feel its heat warming your skin. After baking in the hot sun all day, hard surfaces like buildings, pavements and roads take time to cool down, as they absorb a lot of heat. If you've got kids, you might know what a metal slide feels like in a playground on a hot day. After a string of hot days, sometimes you'll find that it is cooler outside than it is inside your home, especially if you walk through your local park with your dog or sit on the cool grass to picnic with friends. The risk of extreme temperatures is exacerbated in cities because dark, non-permeable surfaces like asphalt and cement and a lack of vegetation result in higher temperatures than surrounding natural areas. For example, the urban heat island effect can add 2°C to local warming, amplifying the impact of heatwaves in cities. Cities in mid-latitude locations of the world are particularly at risk from rising temperatures, where they could be subject to twice the levels of heat stress compared to their rural surroundings under all emissions scenarios. Extreme heat poses serious threats to human health and can cause the failure of critical infrastructure like electricity and transport networks.

As the planet has warmed, we have seen an increase in the frequency, intensity and duration of deadly heatwaves, especially in places like Europe, Asia and Australia. In August 2003, record heat across twelve European countries resulted in at least 70,000 deaths, with close to 15,000 deaths in France alone. Averaged over the continent of Europe, temperatures were up to 5.5°C above the long-term summer average, with France, Germany, Switzerland, Portugal and the United Kingdom sweltering through the hottest European summer in at least 500 years. Daily temperatures across individual countries were 7.5–12.5°C above average. France experienced eight consecutive days with temperatures over 40°C, with the cities of Paris and Toulouse hit hard. In a region not accustomed to very hot summers, many homes are not air conditioned and there was limited public awareness about the need to keep drinking water and stay out of the direct heat. Temperatures remained at record high levels at night, resulting in oppressive conditions that killed thousands of elderly people in hospitals, nursing homes, and their own living rooms. Many bodies were unclaimed for weeks as relatives were away on their summer vacations. The number of deaths was so high that a refrigerated warehouse had to be set up outside of Paris to deal with the overflow of bodies from local morgues.

Aside from the devastating death toll, the heat across the region resulted in heavy crop losses, the melting of 10 percent of Europe's alpine glaciers, and extensive wildfires across Greece, Italy, France Spain and Portugal. The economic losses associated with the extreme heat were estimated to be as high as US$10 billion, with more than 1 billion Euros worth of damage from record-breaking forest fires in Portugal alone. Attribution studies carried out in the aftermath of the event reported that climate change doubled the risk of the 2003 European heatwave, with 70 percent of the deaths in central Paris and 20 percent in London directly attributed to human-caused climate change. These studies were the first to estimate how much human activity increased the risk of a specific weather event and its related death toll, paving the way for impact attribution studies.

The 2003 European heatwave was a stark warning that no part of the world – no matter how wealthy – is immune from the deadly impacts of climate change. The fact that as many as 70,000 people can die in the richest nations on Earth in the twenty-first century shows just how underprepared our society is for the lived experience of climate change. Before the summer of 2003, only Lisbon and Rome had heatwave alert systems. After the event, most European cities adopted extreme temperature prevention and alert plans, which come into effect when extreme weather conditions are forecast. Although the underlying issue of inadequate cooling of homes in areas unaccustomed to extreme heat remains, these new warning systems and increased social awareness following the heatwave will hopefully go a long way to helping people adapt to the reality of their new climate.

Unfortunately, risks to natural ecosystems from extreme wildfire conditions still remain, as there are only so many firefighting resources that can be deployed, and sometimes the capacity to respond is overwhelmed by the sheer enormity and number of blazes that will be ignited in a warmer world. The European summer of 2021 was a graphic example of this, when the Mediterranean countries of Greece, Turkey and Italy were faced with unprecedented wildfires during yet another recording-breaking summer, which was 1°C warmer than the historical average for the continent. We saw shocking scenes of the mass evacuation of holidaymakers in Greece as infernos engulfed tourist destinations like the island of Evia, and fires threatened the southern Peloponnese region near the UNESCO World Heritage site of Ancient Olympia, one of the most sacred archaeological sites in the Western world.

In Athens, the hottest capital city in Europe, people sweltered through the worst heatwave in more than thirty years, causing authorities to close ancient sites like the Acropolis as temperatures reached 42°C. The severity of the event led scientists to begin discussing the need to name heatwaves the same way tropical cyclones and hurricanes are identified, to help people prepare for extreme heat the same way they ready themselves for severe storms. As temperatures soared to 48.8°C in Sicily, setting the highest temperature in European history, the Italian media jumped in, dubbing the blistering heatwave "Lucifer." An attribution study by scientists from the UK Met Office released in November 2021 demonstrated that the record-breaking summer experienced across Europe in 2021 is impossible to explain without the influence of human-caused climate change. They show that Europe can now expect near 50°C heatwaves – which used to be a one-in-10,000-year event in pre-industrial times – to occur every three years as climate change continues to affect weather patterns in the immediate future. If fossil fuel emissions continue along the "middle of the road" emissions scenario we are currently on, Lucifer-grade heatwaves will become an annual occurrence by the end of the century.

The catastrophic summer of 2021 led to the signing of a joint declaration from leaders around the Mediterranean – the area of Europe considered the most vulnerable to climate change – to improve the co-ordination of their disaster response efforts and better integrate climate change policies across the region to ensure the future security and stability of Europe. The Greek prime minister, Kyriakos Mitsotakis, addressed the meeting, saying, "The climate crisis is no longer a distant threat; it has landed firmly on our shores." In the aftermath of the event, the government also appointed Europe's first ever Chief Heat Officer, Eleni Myrivili, to better prepare for the scorching heat that cities like the Greek capital are expected to experience in the coming years.

Although it's a positive sign that the world's wealthiest nations are starting to wake up to the reality of the climate crisis, we are still not anywhere close to responding to the escalating situation like the emergency it is. When you consider the IPCC's finding that about half of the European population may be exposed to high or very high risk of heat stress during summer by 2050 under a high-emissions scenario, not only in the Mediterranean but also in western, central and eastern Europe, it becomes clear that we have an enormous adaptation challenge ahead of us, even in developed countries. The situation is particularly bad in cities, where over 70 percent of Europeans live, where the urban heat island effect will amplify the impact of heatwaves. Highly insulated houses that meet present building standards across Europe will be more vulnerable to overheating unless proper ventilation and other adaptation strategies are actively implemented.

The IPCC estimates that the number of people at high risk of heat-related deaths and heat stress in central and southern Europe will increase two- to three-fold with 3°C compared to 1.5°C warming. Beyond 3°C, the report warns that there are very real limits to the adaptation potential of existing healthcare systems, particularly in the Mediterranean and eastern Europe, where the sector is already under pressure. As we saw during the deadly coronavirus pandemic, hospitals in Italy and Spain were overwhelmed, providing a disturbing glimpse of how climate change will exacerbate underlying social issues that many societies face. If the world fails to rein in the burning of fossil fuels, people fear that Athens could become Europe's first uninhabitable capital city. It's hard to imagine a world where the oldest metropolis in Europe, with its 5000-year history, the very birthplace of democracy, has become too dangerous for people to live in. Climate change is threatening the foundation of human civilization; our own modern-day Greek tragedy is playing out, threatening to destroy life as we know it.

*   *   *

Of all the places I've ever visited, India is the country that left the deepest imprint on me. It's an incomparable mix of the sacred and the surreal, the profound and the downright puzzling; it's the kind of place where the street signs say "accidents are prohibited on this road." After descending into the impossibly thick layer of brown smog that shrouds New Delhi, you leave the air-conditioned shelter of the airport to be hit by a wall of heat, humidity and pollution. The air quality in the world's most polluted city is so bad that it coats the inside of your mouth to the point where you can actually taste it. Black gunk begins to line your nose and lungs in shocking ways. Its toxicity is no joke – heavily polluted air kills close to 2 million Indians each year.

As you walk the frenetic streets, you see the homeless in their thousands sleeping on the street at night: children piled together in railway underpasses, men curled on noisy median strips; humanity jammed into every conceivable space. Kamikaze rickshaws weave through streets choked with a mass of people, cars, cows and bicycles. In the city of Varanasi – the spiritual capital of India where many Hindus go to be released from the endless cycle of death and rebirth – thousands flock to the banks of the holy Ganga. The severely polluted river is a magnet for the maimed and the sick, the dead and the dying. The Ganges at dawn is a sight to behold – people praying, bathing, swimming, shitting, drinking, brushing teeth, washing clothes and burning the dead. It's the full spectrum of human suffering on graphic display. As a Westerner buffered from the reality of life for the world's poor, witnessing the extreme poverty, filth and chaos of India completely reconfigured my world view.

India is the place that my mind often goes to when I think about what a warming planet means for the developing world. In the lead-up to the Indian monsoon, oppressive heat engulfs the region as humidity starts to build during May into the start of June. In May, average temperatures are 35°C in parts of north and central India, but maximum temperatures can soar close to 50°C. Extreme heat can have a devastating impact on human health, resulting in cramps, exhaustion, heat stroke and death. Excessive heat is also known to aggravate pre-existing respiratory conditions, heart disease, kidney disorders and mental illness. In a country like India, chronic air pollution exacerbates many of these problems, with outdoor workers like rickshaw drivers, street vendors and construction laborers, along with children, the elderly and the homeless left highly exposed to dangerous conditions.

In May 2015, northern India was struck by a severe heatwave that pushed temperatures close to 50°C in places, having dramatic impacts on large population centers including the Indian capital of New Delhi. The city endured five consecutive days over 43°C, peaking at 45.5°C on 25 May, causing roads to melt in the extreme heat, distorting road markings as the asphalt liquified. The brutal heat placed a heavy burden on electricity demands as those with air conditioning sheltered in their homes. Outside, exposed people succumbed to extreme dehydration and heat stroke, the lethal heat claiming the lives of more than 2500 people in just a few days. Hospitals were overwhelmed, with staff struggling with the added pressure of power outages as large queues formed outside. Women wrapped their heads in scarves to protect themselves against the blistering sun while they waited for their inconsolable babies to be treated.

According to the Red Cross, heatwave deaths are generally under-reported in developing countries like India. Experts believe that the actual number of deaths from this event is likely to be much higher as it is difficult to source figures from rural areas, and deaths due to pre-existing health conditions that are exacerbated by heat are often not accounted for as they are in places like Europe. The event caught the world's attention as it graphically highlighted the impact climate change will have on highly vulnerable people in poor nations, and the urgent need to prepare society for similar disasters in the future.

Currently around 30 percent of the world's population is exposed to deadly heat for at least twenty days each year. Depending on the emissions pathway the world decides to take, between half and three quarters of the human population could be exposed to periods of life-threatening conditions arising from extreme heat and humidity by the end of the century. While temperate areas will experience more warming than equatorial regions, tropical cities like Lagos, Jakarta and Mexico City will be the worst affected, as these areas have year-round warm temperatures and higher humidity, requiring less warming to exceed "dangerous" heat thresholds. When temperatures increase from 1.5°C to 2°C warming – which could occur as early as the 2040s – an additional 1.7 billion people will be exposed to severe heat. This has major impacts in urban areas in sub-Saharan Africa and in the cities of South and Southeast Asia, disproportionally affecting the poorest people already struggling with intense poverty.

The latest climate modeling presented in the IPCC report indicates that the risk of heat extremes increases with higher levels of warming. Extreme heat events that used to occur once every fifty years in pre-industrial times will be nearly ten times more frequent under 1.5°C of warming, and a staggering forty times more likely at 4°C. Even with 1.5°C of global warming, 40 percent of the largest cities in the world will become heat-stressed, a doubling relative to the 1979–2005 period. South Asian cities will be the most heat-stressed regions of the world in coming years, with population growth in developing regions adding to the challenge. Twice as many people living in megacities like Lagos and Shanghai are expected to become vulnerable, exposing more than 350 million more people to deadly heat by 2050 under a mid-range population growth scenario. India and Pakistan are projected to be particularly at risk, with the IPCC reporting that at 1.5°C, Kolkata will experience heat equivalent to the 2015 Indian heatwave every year; in Karachi, Pakistan's largest city, extreme heat similar to

2015 is expected about once every four years. Under 2°C of warming, both cities can expect conditions equivalent to their deadly 2015 heatwaves every single year.

With 4°C of global warming, heat stress will affect cities in new areas, exposing close to 80 percent of megacities to extreme heat in giant metropolises like Tokyo, Beijing, New York and Rio de Janeiro every year. At such high levels of warming, we start to see very long runs of hot days followed by warm nights that are likely to be very difficult, if not impossible, to adapt to. When you factor in the role of increased humidity in a warmer world, particularly in tropical locations, there is a physical limit to what humans (and other life forms) can adapt to. Research has shown that it's the deadly combination of searing temperatures and oppressive humidity that causes dangerous heat stress. Sometimes referred to as the "silent killer," heat stress occurs when the body absorbs more heat than it can tolerate.

High humidity limits the body's ability to cool itself by sweating, causing the body to overheat. The higher the humidity, the hotter the body feels, so more sweating is needed to cool down. When humidity starts reaching 75 percent, there's already so much moisture saturated in the air that heat loss through evaporative cooling becomes less effective. When the body's core temperature exceeds 40°C, there is serious risk of experiencing heat stroke, a potentially life-threatening condition caused by the body's inability to regulate its core temperature and central nervous system under prolonged exposure to high temperatures, especially in tropical areas.

In places experiencing high levels of heat and humidity for extended periods of time, we can expect to see more heat-related deaths, particularly in concreted cities where warming is already amplified. If we continue along our current intermediate-emissions pathway, South Asia is projected to experience between 84 and 182 days over 35°C each year towards the end of the century (2081–2100), with an average of 128

extremely hot days each year. That's over a third of the year characterized by life-threatening heat. Under a worst-case emissions scenario,
this jumps up to between 99 and 228 days, with an average of 160 days
of extreme heat every year. It's hard to imagine the impact this staggering level of dangerous heat and humidity will have on some of the most
densely populated cities on Earth. These areas already contain slums
filled with the poorest of the poor, who have no way of protecting themselves. Unless we change course, we risk a future where we potentially
face humanitarian disasters every summer across our planet.

*   *   *

Coming to terms with the reality of the world we live in is never easy;
there is a reason why some people live by the proverb "ignorance is
bliss." But one of the upsides of being involved in the IPCC process was
the opportunity to visit amazing countries and expand my world view. It
provided me with firsthand experience of the difficulties faced in different regions. Taking the time to appreciate the very immediate problems
they face was a real perspective hit. It allowed me to look beyond the
shiny veneer, and sense the vulnerabilities that lie just beneath the surface. Not only did it help me realize that no one is safe from the impacts
of climate change, but it also made me realize that there is a lot of collective wisdom that can be gained from considering the world from
another culture's perspective. Our strength lies in our diversity.

In August 2019, our third IPCC lead author meeting was held in
the beautiful riverside city of Toulouse in southern France, one of the
wealthiest regions of the world. As Europe sweltered through one of its
hottest summers on record, a burst of late-season heat took hold, creating an eerily apt backdrop for our gathering. Temperatures soared,
providing us yet another reason to talk about weird weather. As there
was no air conditioning at Météo France, the French national weather

service where the meeting was being held, the organizers provided us with old-fashioned handheld fans to help cool off. We sat there flapping as we debated figure captions, tricky text and how to resolve cross-chapter overlaps in our content.

To make things even more challenging for the non-Europeans, jet lag was warping our brains, a glance at my laptop clock reminding me that my body felt like it was nine p.m. when it was really only lunch-time. We still had hours of technical discussions to get through as sweat trickled down our backs. I've never felt so overloaded as I did at the end of each day of an IPCC meeting. Every discussion required complete concentration and often follow-up action to meet relentless deadlines. Difficult science aside, tuning in to the subtleties of so many foreign accents, while hard even for native English speakers, must have been even harder for many of the multilingual people gathered in the room. We all wilted in the heat until consensus by exhaustion was eventually reached.

At the end of each jam-packed day, we strolled the cobblestone streets in search of a cold drink and somewhere to take in the bustling atmosphere of the European summer. Old men sitting by the elaborate fountain in the town square, children delighting in a carousel from a bygone era, the thrum of cafes packed with families and friends. There is something really comforting about being in a place with such rich human history. The streets are charged with an echo of distant times, collective memories. Centuries of history live on in the spectacular buildings, art and culture all around you. As I wandered the laneways with my weary colleagues, I felt humbled by a sense of something eternal.

Our chapter's French author led us to the architectural wonder of the Basilica of Saint-Sernin, a UNESCO World Heritage–listed church built in 1180. Soaking in the magnificence of the medieval church soothed my over-stimulated mind, shifting me into a deeper, more reflective part of myself. Ancient places have a way of connecting you with something that feels timeless, a reminder that countless generations have come before you. Our ancestors have walked the same streets, grappled with the same inner thoughts, throughout the continuum of history. We exist in a specific moment in time that is ultimately so fleeting, so universal, yet uniquely our own. Ultimately, we are just passing through. Everything around us is the legacy left by everyone that's ever come before. Our world is a testament to the collective choices we've made to create and preserve places of enduring beauty; places that directly return us to our soul.

Being in the ancient city of Toulouse put me in touch with the fact that we are simply the latest generation here to add another layer to human history; the legacy we leave is our choice. As an IPCC author I understand that what we choose now matters more than it ever has before. Stabilizing our Earth's climate is the most important decision humanity will ever face. We are the generation that has the power to make or break our civilization; to be the ones who rose to the greatest moral challenge in human history. We must look deep within ourselves and find our own response to the question posed by American poet Mary Oliver: "What is it you plan to do with your one wild and precious life?"

My answer? Everything I can, with all that I have.

7

Lost worlds

WHERE I LIVE IN SUBTROPICAL eastern Australia, the ocean is crystal clear. In the early morning before the wind picks up, the water looks like glass. Fish dart along geometric ridges that ripple the sand as you wade in. Salt spray refracts over the glistening sea, turning the sky powdery shades of pink, orange and blue. Between the ocean and the land, the coastal heath hides dozens of local and migratory birds that provide hypnotic dawn melodies. After a summer feeding on krill in Antarctic waters, thousands of humpback whales migrate past our shores during winter to breed and give birth in the tropical waters of the Coral Sea. Some populations can travel up to 8000 kilometers each way, making it the longest migration undertaken by any mammal on the planet.

When you take the time to slow down, it's easy to feel the pulse of life all around you. The distinction between the air and your skin is wafer thin; the humidity in the air saturates your pores, sedates you. Sitting by the ocean, you can feel your molecules rearranging, resetting your equilibrium. You dissolve into the landscape. It's a visceral reminder that your body is a part of nature; that nature exists inside of you.

My life wasn't always as blessed as it is now. As I've aged and progressed further into my career, my depression has deepened, leaving me increasingly drawn to live a secluded life far from the cities I've lived and worked in for over forty years. After a few very difficult years spent battling attacks on my work from climate change deniers, we finally made the decision to move closer to my husband's family and enjoy the simple pleasures of sun, sand and surf. When I am really struggling, all the color drains out of my life; everything feels so bleak, so pointless. The world feels sharp against my edges, its sadness piercing open wounds that never seem to heal. In those dark times I fear I will be trapped in a parallel universe of pain from which I can never return. When I am down, it's a battle to bring myself back to the basic needs of my physical body; to breathe, sleep, eat and move.

Since moving to the coast, managing my depression has become easier. It's the best way I've found to keep going in a job that demands so much of me. When I am surrounded by so much beauty, it's easier to distract myself from the more painful realities I face. Other times, when sadness consumes me, all I can do is sit by the sea and cry. When I am away for months teaching in the city, sometimes I lose my bearings and grief overwhelms me:

> I'm starting to feel the pull of home. I need to be by the ocean. It helps calm me down and feel like everything is ok. That life is worth living despite all the heartbreak and horror in the world right now. It feels really hard to keep going like the world isn't crumbling all around us. Part of me just wants to stop dead in my tracks and just scream at all the frustration and sadness that I feel. Some days the growing awareness of just how bad things really are can't be neatly contained in the controlled channels of logic and reason. Sometimes I just want to drop to my knees and howl like a child who just can't understand the injustice of it all. The right thing to do is so obvious and sane that it seems too simple to be true. But it is. Some days I just want to curl up in bed and forget that the world is ending, somehow convince myself that my role in all of this is inconsequential. That everything I am

trying to do isn't so pointless and futile. Today is one of those days. I am just so exhausted and heartsick, for the Earth, and from the impossible nature of the human condition. We all just want to be safe and free enough to enjoy our finite time on this beautiful, battered planet. I feel my heart breaking today.

No matter how grim things get, the ocean is able to absorb all the sorrow I pour into it. In exchange it offers me its wisdom; that life is in a constant state of flux, the ceaseless turn of the tide may be eternal, but my time here is fleeting. When I am in the grip of despair, the ebb and flow of the landscape reminds me that this, too, shall pass. Everything – good or bad – always does, so it's best to hold things lightly. Eventually the pain will transform into a force that melds my broken pieces back together, fortifying a resilient core that somehow still exists inside me. American writer Glennon Doyle writes beautifully of the restorative power of the ocean:

The sound of the water speaks not to my spinning mind or yearning heart, but to my still, strong soul. The water is speaking in a language I knew before the world taught me its language . . . I let it speak to me and I do not speak back. I just receive. I understand with great gratitude that I could rest here forever, offer the sea nothing in return, and it would never stop speaking to me . . . this is not a transaction; it is a gift . . . the surf continues to hit the sand rhythmically and dependably and I trust it will continue. The sun is setting but I know it will rise again tomorrow. There is a pattern to things. This makes me wonder if I can also trust that there is a pattern, a rhythm, a beauty, a natural rise and fall to my life as well. I wonder if the one holding together this sky might also be capable of holding together my heart.

Although my relationship with the ocean has become an integral part of my life, it's nothing compared to the profound connection those who have lived by the sea for countless generations must feel. The ocean is a vital part of the identity in the Blue Pacific, the world's largest oceanic continent made up of a group of island nations that contain the most diverse range of Indigenous cultures on the planet. In the essay "The Ocean in Us," Tongan–Fijian writer Epeli Hauʻofa quotes the I-Kiribati poet Teresia Teaiwa: "We sweat and cry salt water, so we know that the ocean is really in our blood."

Around 11 million people live in the Pacific Islands, an area that is very culturally and physically diverse. For example, there are at least 145 native languages spoken in Vanuatu by around 300,000 people spread across its eighty islands, making it the country with the highest density of languages per capita in the world. On Malekula, the second largest island in the archipelago, more than thirty Indigenous languages are spoken by a population of around 25,000, all distantly related to the Māori language spoken in New Zealand. Geographically, the Pacific contains the rugged volcanic islands of French Polynesia and Hawaiʻi, and the low-lying atolls of the Marshall Islands, Kiribati and Tuvalu that are scarcely above current sea level. The region contains more than 10,000 islands and around 60 percent of the world's coral reefs, which collectively support 25 percent of all marine life. The Pacific Ocean contains the largest tropical and subtropical coral reef habitat on the planet, encompassing a vast 88 million square kilometers from Papua New Guinea to the west coast of Ecuador. It's the tropical heart of our blue planet.

In recent years we have witnessed the devastating impact of tropical cyclones and sea level rise on the Pacific Islands. In February 2016, Tropical Cyclone Winston slammed into the southwest Pacific, severely impacting the nation of Fiji. The category 5 cyclone struck the islands with sustained winds of up to 230 kilometers per hour and a central pressure of 884 hectopascals, making it the most intense tropical cyclone ever recorded in the Southern Hemisphere. Over 60 percent of Fiji's population was affected, with economic losses estimated at US$1.4 billion, wiping out a third of the nation's gross domestic product in a single blow. Close to 1 million people lost power as the ferocious winds downed trees and powerlines across the region. Entire communities were destroyed, with many islands sustaining catastrophic damage, including the destruction of 495 schools and eighty-eight health clinics. Over 30,000 homes – nearly a quarter of all households – were destroyed or damaged by the storm. Winston severely affected the lives of 540,000 Fijians, around 62 percent of the nation's entire population, including the deaths of forty-four people. Over 700 shelters were opened across the affected region, with a state of emergency declared by Prime Minister Frank Bainimarama, who stated that "almost no part of our nation has been left unscarred." The extensive damage left around 150,000 people displaced, forcing them into evacuation shelters including tents and makeshift housing made from tarpaulins.

It is estimated that it takes Pacific Islands around fifteen years on average to fully recover from tropical cyclones. The need to rebuild is psychologically important for those who live through the trauma of an extreme weather event; the ability to imagine a new future is vital for recovery. But in some cases, the land no longer exists, so uprooted people face exile in their own countries. Recovery also depends on whether there is enough time between disasters for people to get back on their feet. In reality, the country has endured a further four category 5 cyclones since Winston hit in 2016, setting back development in the region for decades. The cumulative impact of these disasters has reduced towns to rubble, leaving many people destitute in a nation where a third of the population already lives below the poverty line. And this is just with the 1.2°C of warming the world has experienced so far; it's a terrible sign of what is to come for the Pacific.

While the science of tropical cyclones is dynamically complex and difficult to monitor, the latest IPCC report explains that the average strength of major tropical cyclones has intensified since 1980, and that the power of these storms is projected to increase in a warmer climate. The proportion of cyclones that reach very severe levels of category 4 or 5 systems will increase, along with the amount of heavy rain that falls during these events as a warmer atmosphere is able to hold more moisture. There is also mounting evidence that cyclone zones are shifting poleward as the ocean surface warms, allowing systems to form in new areas that were not historically designed to cope with cyclonic conditions. To make matters worse, rising sea levels will exacerbate the impact of coastal inundation, as huge swells surge further inland, submerging low-lying areas.

The IPCC recognizes that coastal cities and other settlements by the sea are on the frontline of climate change. Global mean sea level has risen faster since 1900 than during any preceding century in at least the last 3000 years, so the risk of coastal hazards will continue to escalate with higher seas. Historically rare extreme sea level events that had a probability of occurring once every 100 years will occur annually by 2100, with some atolls being uninhabitable by 2050. Today, nearly a billion people live on low-elevation coasts directly exposed to climate and other coastal hazards. Around 50 percent of the Pacific's population is estimated to live within 10 kilometers of the coast; at least half of their infrastructure is concentrated within 500 meters of the coast. In the Caribbean, around 22 million people live below 6 meters elevation. Many of the communities in these low-elevation areas are exposed to what the IPCC terms "extreme sea level events": coastal inundation that occurs when water levels increase from changes in sea level, storm surges associated with tropical cyclones, and high tides. Events that were considered rare in the recent past, occurring once every century, are projected to occur at least once a year in many regions by 2050 under all emissions scenarios. Places like the Maldives, Tuvalu, Kiribati, Marshall Islands and the Torres Strait are already experiencing the impacts of sea level rise, where ocean flooding has washed saltwater onto agricultural lands and contaminated sources of drinking water.

Small island nations across the Pacific, the Caribbean and the Indian Ocean, where people are already suffering from the impacts of rising seas, cyclone damage and ongoing freshwater shortages, are particularly exposed to increased flooding, erosion and permanent inundation. The IPCC estimates that with 1.5°C of warming, between 300,000 and 560,000 people in small island nations are at risk of permanent inundation, with the number increasing to up to 640,000 at 2°C of warming, which could occur as soon as the 2040s. To make matters worse, those not forced off their island by higher seas could have their livelihoods stripped away as coral reefs begin to die off as water temperatures rise, even under low greenhouse gas emissions pathways consistent with the Paris Agreement targets. The tourism industry accounts for up to 70 percent of a country's gross domestic product in places like the Cook Islands, as people from all over the world flock to experience the stunning natural beauty of the islands and their rich cultures for special occasions like weddings, honeymoons or the holiday of a lifetime.

The cruel irony is that the people who have contributed the least to global warming are the ones who will suffer the most. In these areas, the sea threatens to engulf the land they live on, endangering the very survival of many island nations. Stabilizing the Earth's temperature as rapidly as possible is literally the difference between life and death for millions of people who will be driven from drowning islands. As Fijian prime minister Frank Bainimarama has said: "We refuse to be the proverbial canaries in the world's coal mine . . . we want more of ourselves than to be helpless songbirds whose demise serves as a warning to others." What the world does now will decide whether our planet's magnificent islands continue to exist or are lost to rising seas. At its core, climate change is a humanitarian issue. Begging the developed nations of the world to allow their cultures to survive is an indignity that no one should be forced to endure. Ignoring their pleas would be the ultimate betrayal, an unforgivable moral failure.

*   *   *

It's three a.m. and I'm awake – again. Since working on the IPCC report, my work as a climate scientist now keeps me up at night. I keep having dreams of being inundated. Huge, monstrous waves bearing down on me in slow motion. Sometimes I stop resisting and allow myself to be sucked out to sea. Perhaps I can survive by surrendering, and just allow the wall of water to consume me. Other times, I watch from the land as an immense tsunami builds offshore. I panic, immediately sensing that I don't stand a chance. I watch the horizon disappear, as I turn to bolt to higher ground. Around me, people are calmly going about their business: friends chatting at outdoor cafes or idly strolling the streets, window shopping.

High water is menacing my subconscious, trying to help me grapple with the overwhelm I feel in my waking life. My stomach burns in pre-dawn darkness, my teeth ache from the nocturnal grinding that my dentist now just acknowledges with a sigh. As each season passes, it's painfully clear to me that we are witnessing the destabilization of the Earth's climate. I worry about what it means for a world already struggling with poverty, disease, war, biodiversity loss, famine and violence. Climate change is often referred to as a "threat multiplier": a force that compounds existing political, social and economic problems that human societies are already struggling to cope with.

The mass displacement of people in areas already living with conflict and insecurity is one of the most devastating consequences of climate change. Every year, millions of people are forced to flee their homes because of conflict and violence, but weather-related disasters are now compounding existing humanitarian crises, undermining the security of millions of people across the world. Since 2008, an average of 20 million people each year have been displaced by weather-related extreme events, with storms and floods ranked as the most disruptive disasters. In 2020, weather-related disasters triggered around 30 million new displacements around the world, with around 70 percent of cases recorded in East Asia, the Pacific and South Asia. According to the Internal Displacement Monitoring Center, the scale of displacement is increasing, with movement occurring within countries' borders as people flee drought-stricken farms, cyclone-damaged villages and areas increasingly submerged by rising sea levels and associated storm surges.

In 2020, the total number of internally displaced people worldwide reached a record 55 million. While 48 million (85 percent) of these displacements were people fleeing conflict and violence in fifty-nine countries including Syria, the Democratic Republic of Congo and Colombia, at least 7 million people across 104 countries were uprooted by natural disasters, mostly the result of tropical storms and floods in nations including China, Bangladesh and the Philippines. But given the logistical issues associated with data collection and reporting in unstable or inaccessible regions during the coronavirus pandemic, the latest figures are likely to be a significant underestimate.

Weather-related disasters were responsible for 98 percent of all disaster displacements recorded in 2020. The main factors behind the record-breaking season were a moderate to strong La Niña event in the tropical Pacific Ocean and warmer than average surface temperatures in the Atlantic Ocean. This resulted in intense cyclones, heavy monsoon rains and flooding in highly exposed and densely populated areas in South Asia, East Asia and the Pacific, including China, the Philippines and Bangladesh. The most active Atlantic hurricane season on record hit the Americas hard, resulting in 2.8 million new displacements across seventeen countries including the United States, Guatemala, Honduras and Nicaragua, where highly destructive Hurricanes Laura, Eta and Iota caused major damage. Meanwhile, extended rainy seasons across the Middle East and sub-Saharan Africa uprooted millions more.

Interestingly, in 2020 the United States had the fifth-highest number of new displacements by a natural disaster, with 1.7 million people forced to flee their homes due to the impact of hurricanes and wildfires in places including California, Louisiana and Texas. In late August, Hurricane Laura made landfall in southwestern Louisiana, a category 4 system with winds reaching up to 240 kilometers per hour, making it the most powerful hurricane to hit the state since 1856. A 4-meter storm surge caused severe destruction along the coast, extending as far inland as the city of Lake Charles, where homes were submerged in muddy floodwaters. More than 585,000 people were ordered to evacuate in Louisiana and Texas, with at least 22,000 people fleeing to emergency shelters when the storm hit. There was extensive damage to buildings, forcing people whose homes were made unlivable into temporary accommodation including hotels and dormitories. The damage bill was estimated to be around US$19 billion, with places like Lake Charles still struggling to rebuild over a year after the event, with around 3000–5000 people permanently displaced by the hurricane.

While these numbers are high by developed-nation standards, they pale in comparison to impacts experienced in the developing world. In 2020, the summer monsoon was particularly strong, with flooding triggering more than 5.8 million new displacements in densely populated regions of Asia, with 5.1 million people impacted in China alone. As the planet has warmed, we have seen an increase in the frequency and intensity of heavy rainfall and flooding. Extreme rainfall events are becoming more likely because a warmer atmosphere can hold more moisture, about 7 percent more for each additional degree of temperature, which makes wet seasons and heavy rainfall events even wetter. The flooding in China affected more than 63 million people across twenty-seven provinces and left more than 200 people dead or missing, causing economic losses of around US$25 billion. Flood heights reached record levels in seventy-seven rivers, causing over 4 million people to be evacuated, with nearly 400,000 homes destroyed by floodwaters. Then, again, in July 2021, close to a year's worth of rain fell in just three days in the central province of Henan, resulting in catastrophic flooding that affected close to 14 million people. Authorities say the rains displaced more than a million people, claiming at least 300 lives. If this is the type of extreme rainfall we can experience with just over 1°C of warming, it's shocking to think about the impacts the world will experience under higher levels of global warming.

*   *   *

While individual weather events have devastating consequences for people all over the world, it's fundamental, long-term changes in climate that will have the biggest influence on our societies. Perhaps the most graphic example is the permanent displacement of people from increases in sea level that are already committed and will play out over coming centuries. Global sea level rise is projected to accelerate during the coming decades, likely increasing between 30 centimeters and 1.1 meters by the end of the century. Estimates of the global population at risk of displacement due to sea level rise vary from tens of millions to hundreds of millions of people depending on the emissions scenario, land elevation baselines and future population growth. Some of the latest research shows that if sea level reaches 1 meter above 2020 levels by the end of the century, at least 410 million people will be at risk of inundation in areas less than 2 meters above mean sea level, particularly in major river delta regions in tropical Asia. Note that this estimate assumes no increase in population numbers from 2020 levels, so the number is likely to be conservative.

In contrast, the IPCC's special report on *Global Warming of 1.5°C* cites research that estimates that the coastal population living below 10 meters is projected to reach between 830 million and 1.2 billion by the end of this century. The latest IPCC figures published in 2022 suggest that the situation might be even more dire than first thought, with approximately 1 billion people living in low-lying cities and other settlements on the coast and on small islands projected to be at risk from sea level rise and storm surges as early as 2050 rather than by the end of the century. This situation is expected to worsen under high-emissions scenarios.

Recent extremes are already providing a preview of the vulnerability faced by low-lying nations of the world. In 2020 in Bangladesh, where two thirds of the country is located less than 5 meters above sea level, the longest monsoon season since 1988 submerged around a quarter of the country. Some 5.4 million people were impacted as the event peaked in August 2020. The flooding triggered around 1.9 million displacements, with people seeking refuge in government shelters, along roadsides and on elevated embankments. Sea level rise alone could permanently displace up to 1 million people in southern Bangladesh by direct inundation by 2050, increasing internal migration six-fold between now and the middle of the century. By 2100, the figure rises to over 2 million people, placing heavy pressure on housing, jobs and food demand in destination locations like Dhaka. The enormous scale of recently observed impacts highlights the vulnerability faced by many developing nations where climate change will worsen living conditions for people already struggling to make ends meet.

In parts of the Pacific, some islands have already been lost to rising seas. In 2016, a study led by Simon Albert from the University of

Queensland reported that five Solomon Islands have been inundated since 2014, destroying villages that have existed since at least 1935. Another six islands had shrunk by 20 to 62 percent, confirming anecdotal reports of people living in the area who have witnessed houses wash into the sea since 2011. Similar work carried out in Micronesia in the northwestern Pacific, where sea level has been rising faster than the global average since 1950, has shown that some islands have now disappeared. Local people told researchers that two former islands have vanished completely, including Nahlapenlohd, an island famous for hosting a great battle between warring chiefdoms in 1850. Aerial images have also revealed that another six low-lying islands in the unpopulated Laiap, Nahtik and Ros island chains became submerged between 2007 and 2014, a sign that things are on the move.

As sea levels continue to rise, many people in the Pacific are being forced to move to higher ground. This is already happening in the Carteret Islands of Papua New Guinea, located just 1.2 meters above sea level, where a resettlement scheme is underway to move the population to Bougainville, a more elevated island 85 kilometers away. Since 1994, the atolls have already lost about 50 percent of their land, resulting in water and food shortages that have forced 1700 people from their homes. Encroaching saltwater has contaminated freshwater wells and turned vegetable plots into swampy breeding grounds for malaria-carrying mosquitoes. The salty conditions now prevent people from growing taro, the staple food crop, on the island. In 2005, their plight gained international media attention, as the Carteret Islanders became known as the world's first "climate refugees."

Like migration anywhere, simply transplanting people into new communities is not always straightforward; there are often significant cultural, social and political differences between displaced people and their host community. Adding to the difficult situation is the challenge of finding land in the Pacific Ocean. Not only is it limited, but it is covered by customary land ownership. For example, 97 percent of the land in Papua New Guinea can't be bought or sold – the land belongs to groups rather than individuals. Relocated people from the Carteret Islands are struggling to sustain their livelihoods due to the lack of available land and are suffering from a loss of their traditional culture.

While permanent relocation might be the only option for some, most Pacific Islanders would much prefer to stay in their homelands and instead receive help to adapt to climate change wherever possible. For many, the far more devastating loss is not of their physical home but that of their traditional culture. In Pacific Island countries, land is critically important to cultural and spiritual identities that have been passed down from generation to generation. In many island nations, the term for land is the same as the generic term for people from that place – they are considered inseparable. In fact, most Polynesian terms for "land" are the same as or similar to the word for "placenta," high-lighting the intimate link between people and their ancestral grounds.

This is the case in Samoa, where a new baby's "fanua," or umbilical cord, is often buried under a tree in the village; people are literally a part of the island they come from. This rooted sense of identity allows people to temporarily migrate with the knowledge that they will always have a home to return to. But if a community is forced to relocate from its homeland permanently, this critical connection with their ancestors is lost. As people are forced to leave, they are considering the mammoth job of relocating entire communities, including graves and the bones of their dead, as Pacific cultures believe ancestors belong with the living.

Losing their connection with their land is just about the worst thing that could happen to a Pacific Islander. Some people would rather die than leave their birthplace. The profound loss of identity, culture and family ties that comes about from leaving ancestral homelands is further compounded by potential conflict and social issues that arise where resettlement occurs in other groups' customary land. In Fiji, six villages have already been moved because of sea level rise, with a further forty earmarked for future relocation. While moving to closely related family areas has reduced the cultural impact in some parts of Vanua Levu, there are reports that moving some coastal communities inland has impacted villagers' spiritual ties to the ocean. In other areas, people with very strong attachments to their land would rather face the risk of climate change than lose their cultural identity and direct connection with their ancestors.

The region is still scrambling to come to terms with this new reality and the cultural and legal complexities of relocating displaced people in an increasingly volatile world. There are calls for wealthier Pacific rim countries like Australia, New Zealand and the United States to do more to help, but the number of people allowed through formal agreements remains low and politically sensitive. Given that the UN Human Rights Committee only recognized climate change–related displacement as a global refugee issue in 2020, it remains unclear how nations will navigate this brave new world.

We are already seeing people forced from their homes because of the slow creep of sea level rise. In the low-lying areas of the Pacific, where entire nations could face extinction within years or decades, ancient cultures face the impossible choice of being uprooted from their spiritual homelands or moving to higher ground. There are so many ethical challenges that are only just starting to be teased out. The question is, how much worse are we prepared to let things get before we truly appreciate the scale of what is at stake? In her powerful speech to the United Nations COP26 climate summit in November 2021, the prime minister of Barbados, Mia Mottley, called out the lack of political ambition to keep the 1.5°C target within reach as being rooted in greed and selfishness, saying "this is immoral, and it is unjust." She went on to challenge her fellow world leaders: "Are we so blinded and hardened that we can no longer appreciate the cries of humanity?," calling on all "those who have a heart to feel" not to commit small island nations to a death sentence.

If we continue on as we are, we will consign the rich cultures of the Pacific and the Caribbean to anthropology textbooks, something children will only learn about in history class. If we lose sight of our shared value of protecting the fundamental human rights of our most vulnerable people, we will stray dangerously far from our inherent humanity. If we abandon our moral compass, in the end – no matter where we live – we will all become lost people in a lost world.

8

―――――

A thousand generations

I'M SURE BY NOW, MANY of you want to put this book down and stop reading. Trust me, I understand how you feel. These last few chapters have been really, really difficult for me to write. Like everyone, I've been trying to keep my head above water during the coronavirus lockdowns that have warped everyone's sense of normality. I'm also writing in the aftermath of completing the IPCC report while simultaneously teaching two university climatology courses and trying to keep my own research alive, so I am the most burnt out I've ever felt in my entire life. I've found myself overcome by tears many times as I've come to terms with the reality of what I'm writing, especially material that I don't deal with directly as a climate scientist. Usually I'm working with physical variables like temperature and rainfall that can be neatly analyzed and understood. But when you start to understand the reality of what the numbers actually mean for the people and places we love, you find yourself face to face with something so profoundly sad. Sometimes in life, there is nowhere to hide from the terrible truth.

The last time I felt this way, I was visiting my father in hospital following emergency surgery for a massive brain hemorrhage. As he lay unconscious in intensive care, I examined his CT scan with one of the attending surgeons, who gently explained that the dark patch covering nearly a quarter of the image of his brain was a pool of blood. Although they had done their best to drain the area and stem the bleeding, the catastrophic nature of the damage was undeniable. The brutality of the evidence was clear – the full weight of it sent my stomach into free-fall. All we could do was watch, as life extinguished itself in agonizing fits and starts.

Right now, the urge to stop writing this book is very real: I want to protect myself from the pain that simply shatters me. This process has put me in touch with a sense of grief that sometimes leaves me feeling like a broken mess. I have spent hundreds of hours trawling through countless UN reports and scientific papers until my eyes sting and I can no longer absorb any more information. I feel overwhelmed and saturated with sorrow. As a sensitive person with a difficult background, sometimes I find the reality of the world we live in unbearable. I just can't understand why we inflict so much pain on each other and our planet. There are days when it's hard to take in all of the senseless destruction and somehow try and accept so much avoidable suffering. Increasingly I fantasize about quitting my job; I dream of living a simple, escapist life by the sea. I want to learn how to be delusional, to be somehow blind to the reality of the world shifting before my eyes.

The problem is, I know that this book must be written, and fast, because far too much is at stake. While I struggle to not give in to my own despair, I know it's wrong to expect young people and the unborn to clean up the world's mess. They have inherited a problem not of their own making. My privilege, education and conscience deny me the luxury of looking away. Although sometimes it takes a heavy toll, I want to share everything I can as quickly as possible while I still can because I know that time is running out. I've come to realize that the only way forward is not a detour through denial, but straight through the heart-land of grief. When you realize that all that sustains us is at stake, it's almost too much to process. The reptilian brain wants to take flight and avoid confronting the unthinkable danger bearing down upon us.

The poet T.S. Eliot was onto something when he wrote, "humankind cannot bear very much reality."

To shy away from difficult emotions is a very natural part of the human condition. But just because we can't face something doesn't mean it disappears. Blocking feelings of empathy and concern to avoid psychological pain is a common defense mechanism designed to protect us from becoming too emotionally overwhelmed. Endlessly distracting ourselves with mundane matters is a way of distancing ourselves from feeling conflicted and distressed by the realization that we, individually and collectively, have an ethical dilemma to face around caring about each other, and the future of all life on Earth. Unfortunately, we live in a culture where we actively avoid talking about hard realities; darker parts of our psyche are considered dysfunctional or socially unacceptable to share. Especially in public. But trying to be relentlessly cheerful, stoic or avoidant in the face of serious loss simply buries more authentic emotions that will eventually come up for air. Denial only ever rests in a shallow grave.

When we talk about climate change, people are often nervous to acknowledge the painful feelings that accompany a serious loss. We quickly skirt around complex emotions, landing on the safer ground of practical solutions like signing up for renewable electricity to feel a sense of control in the face of far bleaker realities. We are afraid to have the tough conversations that connect us with the darker shades of human emotion and shameful collective histories. But grief is not something to be pushed away: it is a sign of the depth of the attachment we feel for something, be it a loved one or the planet. If we don't allow ourselves to grieve, we stop ourselves from emotionally processing the reality of our loss. It prevents us facing the need to change the way we live and respond to our new reality with an open heart.

As more psychologists begin engaging with this topic, they are telling us that being willing to acknowledge our personal and collective grief might be our only way out of the planetary mess we are in. When we are finally willing to accept feelings of intense loss – for ourselves, the planet, and every child's future – we can use the intensity of our emotional response to finally propel us into action. We must have the heart and the courage to be moved by what we see. Because the truth is that life as we know it hangs in the balance; every fraction of a degree of warming matters. Every year of further delay matters. It's the difference between how much we destabilize the ice sheets, the amount of dangerous heat we are exposed to each summer, and whether or not millions of people lose their homes to rising seas. The longer we delay, the more irreversible climate change we will lock in. Any young person can tell you that stabilizing the Earth's climate is literally a matter of life or death. It will impact the stability of their daily lives, their decision to start families, and their chance to witness the natural wonders of the world as their parents did. The ability of current and future generations to live on a stable planet rests on the decisions the world collectively makes right now.

So to all the young people and parents reading this book: I keep writing this because of you. I want you to know that there are scientists who really care about the future we are leaving our kids and their kids a thousand generations from now. We did all that we could to warn the world. We tried to minimize the intergenerational damage. What happens next is up to all of us: who you vote for, what you buy, how you spend your brief time on this planet matter more than ever before. So next up is my rundown of what you need to know about what climate change means for your life and generations to come. From there, we can choose what happens next.

*   *   *

The first thing to know is that climate change is not fair. It is something that the rich nations of the world have caused through the burning of

fossil fuels and clearing of the land, all in the name of economic progress. The world's dominant economic model is capitalism, which rests
on the exploitation of the planet's natural resources and the poor for
corporate profit, often with scant regard for the collective good or the
wisdom of First Nations peoples. Since the end of World War II, capitalism has turned humans into consumers and the Earth into a giant
quarry to generate wealth for people to live comfortable lifestyles, predominantly in rich nations. Before the Industrial Revolution, humans
were largely sustained by small-scale agriculture that supported local
communities. People bought and sold what they could make, bake,
sew or grow.

Industrialization saw societies shift to the large-scale use of technology and machinery to increase the rate and scale of the production
of goods for sale in competitive global markets. To keep production
costs down, manufacturing has often been located in developing countries like China, Bangladesh and Mexico where impoverished workers
are desperate enough to work for negligible wages to feed their families. As mechanization and the development of fertilizers exploded,
the humble family farm gave way to industrial-scale agriculture on
land that once supported natural ecosystems. As the size of the human
population grew exponentially, the scale of mass production exploded
to meet the insatiable needs of human societies, now influenced by
advertising targeted to manipulate people into buying things they
really don't need. The wealthier a society becomes, the more it consumes. For example, in just over two days, the average American emits
as much as an average Nigerian does in an entire year. The wealthiest
10 percent of people on the planet own three quarters of all economic
wealth, while the poorest half of the world's population owns just
2 percent. Economic power – and its associated political influence – is
concentrated in the hands of a very small minority of the super-rich, at
the expense of the world's poor and nature.

All of this industrial activity – from the manufacturing of smart phones to intensively farmed cattle ranches – has been driven by burning fossil fuels and clearing the land, which has altered the energy balance of the planet. While developing nations have also contributed to heat-trapping greenhouse gas emissions in recent decades, their exposure to the impacts of climate change far outweighs their contribution to the problem. When ranked by income, the fifty wealthiest nations are responsible for 86 percent of cumulative carbon dioxide emissions. The poorest fifty nations are responsible for just 14 percent. North America and Europe alone are responsible for 40 percent of all emissions released between 1850 and 2019. Today, China, the United States and the European Union are the three largest greenhouse gas emitters, with the United States having the highest emissions per person. A study by Oxfam and the Stockholm Environment Institute reported that between 1990 and 2015 the richest 1 percent of the world's population were responsible for more than twice the carbon emissions generated by the poorest half of humanity. The poorest 50 percent were responsible for just 7 percent of cumulative emissions, highlighting the shameful social and economic inequalities that have led to the planetary emergency we now find ourselves in.

While there is a certain level of committed warming locked in until we mop up the historical emissions released since the Industrial Revolution, how bad we let things get is still in our hands. The IPCC report explains that once we begin to rapidly remove greenhouse gases from the atmosphere and reach net-zero emissions, the Earth's temperature will begin to restabilize and stop warming. While that is good news, unfortunately a zero-emissions world does not mean that climate change impacts also stop. This is because melting glaciers and ice sheets and associated sea level rise all occur slowly and lag behind surface temperature warming. This means that even after we have put the brakes on emissions, sea level will continue to rise over hundreds and thousands of years due to continuing deep ocean warming and ice sheet melt and will remain high for millennia. The latest IPCC figures tell us that the world will be committed to 2–3 meters of global sea level rise if warming is limited to 1.5°C, rising to between 2 and 6 meters at 2°C over the next 2000 years.

The IPCC's *Sixth Assessment Report* clearly shows that every fraction of a degree of warming matters and that the impacts of climate change are not distributed evenly around the world. In particular, the children of low- and middle-income countries have borne the brunt of losses to land, cultural heritage, Indigenous knowledge, biodiversity and health as a result of climate change. If we don't change our ways, children in developing nations will be burdened with the most dangerous impacts of the climate crisis. Under current policy pledges, children born in 2020 will experience up to seven times more extreme weather and climate events, especially heatwaves, compared to people born in 1960. The amount of land area affected by heatwaves will increase from 15 percent today to 22 percent by 2100 if we limit warming to the Paris Agreement goal of 1.5°C. Current policies see that number jump to 46 percent of global land area experiencing heatwaves by the end of the century, with the Middle East and North Africa identified as the most vulnerable areas.

With 3°C of global warming – a threshold breached in all but the two lowest future emissions scenarios – a six-year-old in 2020 will be exposed to twice as many wildfires and tropical cyclones, three times more floods, four times more crop failures, five times more droughts and thirty-six times more heatwaves than a child born in 1960. Behind the global average, there are important differences that exist from region to region. Children born now and in the future are much more likely to be in low-income countries, so people in these areas will be disproportionately exposed to weather and climate extremes. For example, under current emissions pledges, by the end of this century children born in Europe and Central Asia between 2015 and 2020 will experience around four times more extreme events than experienced in these areas today. Meanwhile, in sub-Saharan Africa, children of the same age will be exposed to up to six times more extreme events, including around a fifty-fold increase in heatwave exposure. Limiting global warming to 1.5°C, instead of following the current intermediate-emissions scenario, nearly halves the additional exposure of newborns to extreme heat, while reducing the risk of crop failures, droughts and river floods by over a third.

While it is true that the human spirit is resilient, there are only so many knocks a person can take. As global warming worsens, exposure to weather and climate extremes also increases, with dangerous conditions eventually forcing people from their homes. This will sever long-held ties to ancestral lands, leaving younger generations disconnected from their traditional livelihoods and culture. The World Bank estimates that without concrete climate adaptation and sustainable development plans, more than 216 million people could be forced to migrate within their own countries by 2050. The top "hotspots" of internal climate change migration are sub-Saharan Africa, East Asia and the Pacific, and South Asia, which collectively account for 80 percent of projected displacements. Millions are also expected to be uprooted from North Africa, Latin America, Eastern Europe and Central Asia, with water availability and declining crop productivity driving many from their homes.

What mass displacement means for the social cohesion and stability of people living in these areas is yet to be seen. Emerging research shows that as young people become increasingly aware of the current and future global threats of climate change, their mental health starts to suffer. A groundbreaking study, led by Caroline Hickman from the University of Bath in the United Kingdom, conducted the largest international survey of climate anxiety in young people and their perception of governments' responses to climate change compiled to date. Their research collected data from 10,000 young people aged sixteen to twenty-five years across ten countries spanning the developing and developed world, from a range of social and economic backgrounds. Their results were damning, but probably no surprise to anybody paying close attention to the people in their life.

Many young people from all over the world reported significant emotional distress related to climate change: they feel that they have no future, that humanity is doomed, and that governments are not doing enough to address the crisis. They feel betrayed and abandoned by adults and governments who have failed to care for the planet and protect their future. They experience a wide range of complex emotions including sadness, anger, powerlessness, shame, despair, depression and grief. Many reported pessimistic beliefs about the future, fearing that they won't have access to the same opportunities their parents had, and that the things they value will be destroyed. Many also feel hesitant to have children under such bleak circumstances. Distress levels are highest when people believe that government responses are inadequate.

As many trauma survivors will tell you, it's often the lack of appropriate response from authorities in the face of traumatic events, rather than the experience itself, that causes the most psychological damage. When there is no acknowledgment of the damage that has been done, no moral consequences for turning a blind eye, it's like it never happened. How can young people establish trust in the very institutions that allow perpetrators of intergenerational damage to roam free? How do they live with the knowledge that the people who are meant to keep them safe are the very ones allowing the criminal destruction of our planet to continue?

Hickman and colleagues describe the behavior of adults and governments as contributing to what British psychologist Sally Weintrobe terms a "culture of uncare." They argue that this, in turn, inflicts significant moral injury on young people as the powerful fail to uphold fundamental moral beliefs about care, compassion, planetary health and ecological belonging. As one of their interviewees shared: "I don't want to die. But I don't want to live in a world that doesn't care about children and animals." The researchers argue that subjecting young people to climate anxiety and failing to act can be regarded as cruel, inhumane, degrading and even torturous. It provides insight into the current phenomenon of children and young people voicing their concerns through legal cases as an attempt to have their distress recognized and validated in the face of government inaction. While these individual efforts are an important way for young people to try and do what they can to address injustice, it is vital that action is taken by those in power to respond to the distress and concerns of younger generations. In her blistering speech at the 2019 United Nations Climate Action Summit, Swedish climate activist Greta Thunberg channeled the sense of anger and betrayal that so many young people feel all over the world:

> This is all wrong. I shouldn't be up here. I should be back in school on the other side of the ocean. Yet, you all come to us young people for hope. How dare you! You have stolen my dreams and my childhood with your empty words and yet I'm one of the lucky ones. People are suffering. People are dying. Entire ecosystems are collapsing. We are in the beginning of a mass extinction and all you can talk about is money and fairy tales of eternal economic growth. How dare you! For more than 30 years, the science has been crystal clear. How dare you continue to look away and come here saying that you're doing enough when the politics and solutions needed are still nowhere in sight. You say you hear us and that you understand the urgency, but no matter how sad and angry I am, I do not want to believe that. Because if you really understood the situation and still kept on failing to act then you would be evil and that I refuse to believe . . . You are failing us, but the young people are starting to understand your betrayal. The eyes of all future generations are upon you and if you choose to fail us, I say:

We will never forgive you. We will not let you get away with this. Right here, right now is where we draw the line. The world is waking up and change is coming, whether you like it or not.

Listening to Greta's devastating speech still sends chills down my spine and tears to my eyes. Young people have every right to feel white-hot fury about the mess we are in. Their future is being sabotaged by the very people who should be protecting them. As the UN Secretary-General, António Guterres, told the World Leaders Summit at COP26 in November 2021:

> Our addiction to fossil fuels is pushing humanity to the brink. We face a stark choice: either we stop it or it stops us. It's time to say: enough. Enough of brutalizing biodiversity. Enough of killing ourselves with carbon. Enough of treating nature like a toilet. Enough of burning and drilling and mining our way deeper. We are digging our own graves . . . Young people know it. Every country sees it. Small Island Developing States – and other vulnerable ones – live it. For them, failure is not an option. Failure is a death sentence . . . On behalf of this and future generations, I urge you: choose ambition. Choose solidarity. Choose to safeguard our future and save humanity.

We know exactly what we need to do, but we still aren't prepared to do it. Instead, we watch extreme weather increasingly ravage every corner of the world with every passing season. Right now, even following the United Nations' 26th Conference of Parties (COP26) in Glasgow, emission pledges are still not on track to achieve the Paris Agreement target of limiting global warming to 1.5°C above pre-industrial levels. Considered by many as humanity's "last chance" to stabilize the climate, instead current net-zero emissions pledges have us hurtling towards global warming of 1.4–2.8°C. And that's a best-case scenario, only if all commitments – which are not legally binding, or in the case of developing nations, not yet adequately financed – are honored completely. Currently implemented policies have us tracking 1.9–3.7°C of warming by the end of the century, with a best estimate of 2.6°C. As you know now, this level of global warming will reconfigure life as we know it.

While I desperately hoped that COP26 would be the political tipping point that changed everything, my rational mind knew that some governments – like my own here in Australia – are still in the stranglehold of the fossil fuel industry. There are corporate interests that are willing to sacrifice our planetary life-support system to keep the fossil fuel industry alive for as long as humanly possible, using unproven technology. Carbon capture and storage, known as CCS, is based on the idea that you can extract carbon dioxide from the smokestacks of coal plants or steel factories, compress it, transport it and then inject it back underground, where, in theory, it will remain forever. And that's assuming you can find the right geologic conditions that are stable enough over millennia so that carbon doesn't leak out and back into the atmosphere.

The problem is not only that the technology is enormously expensive, but that despite over twenty years of research, it is still unproven to work at the scale required to substantially reduce emissions. According to the Global Carbon Capture and Storage Institute, there are twenty-seven operational CCS facilities globally, predominantly in the United States, jointly able to capture 36.6 million tonnes of carbon dioxide annually. For context, the world emitted 39.4 *billion* tonnes of carbon dioxide in 2021 – that's roughly 1000 times greater than what's possible to capture with current CCS technology. Put another way, CCS plants can only offset around 0.1 percent of global carbon emissions each year. To reach net-zero emissions by 2050, scientists calculate that carbon dioxide needs to decline by approximately 1.4 billion tonnes each year. The industry group estimates that between US$655 billion and

US$1280 billion is required to make this a reality. Aside from the trillion-dollar price tag, it's critical to realize that CCS projects take around ten years to progress through concept, feasibility, design and construction phases before becoming operational – time we simply don't have.

Relying on technology that is not ready to be deployed on the scale needed to immediately and drastically address the emergency we face is at best reckless, and at worst an intergenerational crime. It also delays facing the reality that we must stop burning fossil fuels – we need to take serious action and not rely on unproven technology to save the day. As people in climate justice circles like to say, "delay is the new denial." We need to turn the tap off new carbon emissions and start mopping up the damage.

While addressing climate change is highly complicated, the truth is simply this: we must leave fossil fuels in the ground to stabilize the Earth's climate. A 2021 study led by Dan Welsby from the University College London calculated that around 90 percent of coal and nearly 60 percent of oil and natural gas reserves need to remain unextracted to give the world a 50 percent chance of stabilizing global warming at 1.5°C. The authors conclude: "This implies that most regions must reach peak production now or during the next decade, rendering many operational and planned fossil fuel projects unviable." Despite the IPCC also clearly demonstrating that virtually all of the observed warming to date has been driven by greenhouse gas emissions (namely carbon dioxide), even nations such as the United States, Germany and the United Kingdom, which claim to be leading the transition to clean energy, are still planning to develop new coal mines, oil fields and gas reserves.

While more than forty countries committed to phasing out coal-fired power before the middle of the century at the COP26 summit in 2021, some of the world's biggest coal-dependent nations, like Australia, China, India and the United States, did not sign up. As long as the protection of the fossil fuel industry continues, real reductions in greenhouse gas emissions will not be possible in time to avert disaster. It is clear that many of our political leaders still don't have the heart or the courage to be moved by the tragedies we see unfolding all around us. Despite the overwhelming evidence of an escalating crisis, they are choosing to continue to put profits over people and the planet. These are the people we have voted for, the people we have put in charge of our future.

As a climate scientist, I find myself in the extraordinary position of trying to write about an emergency that is unfolding in real time; I don't know how this story will end. So in solidarity with young people all over the world – the millions of School Strike for Climate protesters and countless other climate action groups fighting for your future right across the planet – I want you to know that we are listening; we will not leave you to clean up this mess alone. Many scientists also feel angry, scared and sad. We don't often talk about it openly, but that doesn't mean we don't feel really upset about what's going on. So just for the record, here's how I felt watching the Northern Hemisphere's brutal summer of 2021 play out in the final weeks leading up to the release of our IPCC report:

Sometimes it's so overwhelming to realize how fast the world is changing. I'm working on my book at the moment and it's making me feel so damn sad. I was writing about the recent [June 2021] Canadian heatwave and realized that there are now so many places that are being fundamentally transformed. In all likelihood I won't ever get to experience those places as they once were. So many places have become so disturbed and degraded. It feels like things are starting not only to crumble, but come away in huge chunks, disintegrating before our eyes. These days I'm finding it hard to bear witness to the increasing instability. Even the most conservative scientists I know of are starting to share their own sense of panic on Twitter. It's a recent shift that confirms my suspicion that something really terrible is unfolding, something that creates a sickening sense of dread in the pit of my stomach. It makes me want to recoil to protect myself from the horror of it all. It breaks my heart that we can sit back and allow the abuse of our planet to continue so unashamedly. If our planet were a child, there would be a moral outcry of disgust and rage – how can we bear to stand back and watch the life be beaten out of the very thing that sustains us? It's brutal, horrific, to think about how badly we have abused our Earth. I feel my heart breaking today.

We all have our dark moments; it's a rational response to a really distressing situation. Sometimes the best thing to do is just let the sadness flow through you; it's okay to feel the way that you do. As a wise person once told me: it's okay to cry. The thing to know is that eventually these feelings will pass, like storm clouds passing through the sky. Everyone has good and bad days; when you feel overwhelmed, it's important to take a break to regain your perspective and remember the bigger picture. It's easy to focus on the people making things worse, overlooking all of the incredible people doing everything they can to make the world a better place.

You don't have to look far to see there is still so much goodness in humanity. Think of our healthcare workers on the frontline of the COVID-19 crisis, volunteer firefighters protecting our precious places, protesters from all walks of life taking to the streets to give future generations and nature a fighting chance. IPCC scientists from all over the world who worked thousands of unpaid hours through a deadly pandemic to produce the most comprehensive climate report humanity has ever compiled. We all do these things because we care about each other and our planet. As I've been writing this book, I have a quote by Indian lawyer and social activist Mahatma Gandhi right here on my desk that says: "In the midst of death, life persists; in the midst of untruth, truth persists; in the midst of darkness, light persists." It's a vital reminder that, no matter how bad things get, in all darkness, there is light.

While the scale of the problem is overwhelming, don't let the fact that you can't do everything stop you from doing something. What you choose to do, no matter how small, makes a difference, even if you can't see how right now. You can be part of the generation that chooses a better future for everyone. In her book *Believers: Making a Life at the End of the World*, American poet Lisa Wells writes:

> If our descendants are alive and well in a hundred years, it will not be because we exported our unexamined lives to another planet; it will be because we were, in this era, able to articulate visions of life on Earth that did not result in their destruction ... Formerly my idea of sustainability had been vague and of the "leave no trace" variety ... ways of life are possible in which human beings not only thrive but also repair damage and even increase biodiversity and beauty of the planet. It is a story predicated on leaving a trace, a legacy.

I hope that enough people like you are waking up right across the planet and saying: enough is enough – we demand a better world. The science is telling us that this is our last chance to avert planetary disaster. Let's choose to leave a legacy of care and repair, one that sees humanity not only survive but thrive. The good news is that the revolution we have all been waiting for is already happening. We just don't hear much about it, as the fossil fuel industry has run a relentless fear campaign to protect their corporate interests. The truth is that change is already sweeping the world, from boardrooms to homes right across the planet.

This is humanity's moment to right the wrongs of the past, to heal our relationship with each other and all life on Earth. It's now time to meet the visionaries who are already showing us that another world is not only possible, but inevitable. And unstoppable.

A new day is dawning.

PART 3

The Whole

9

In all darkness, there is light

POLITICAL RECOGNITION OF THE THREAT of climate change goes back at least thirty years. Acknowledgment that human activity can dangerously influence the Earth's climate can be traced back to the United Nations Conference on Environment and Development held in Rio de Janeiro, Brazil, in 1992. Commonly referred to as the Rio Earth Summit, the landmark meeting resulted in a number of significant environmental outcomes, including the first international climate treaty, the United Nations Framework Convention on Climate Change (UNFCCC), which came into force in 1994 with the goal of avoiding "dangerous human interference with the climate system."

Since 1995, there have been yearly meetings of the Conference of Parties (COP) signed up to the UNFCCC to assess global progress in dealing with climate change. In 1997, COP3 was held in Kyoto, Japan, where after intensive negotiations, the Kyoto Protocol was adopted to reduce greenhouse gas emissions. The agreement acknowledged that different countries have unequal capacity to deal with climate change based on their development status. On the basis that developed countries are historically responsible for the current levels of greenhouse gases in the atmosphere, it was agreed that richer nations are obliged to reduce greenhouse gas emissions, with developing nations only urged to contribute voluntarily to the effort. More than 100 developing nations, including China and India, were exempted from the agreement altogether. The exclusion of these two rapidly growing economies – accounting for over a third of the world's population – was used by critics in developed nations like the United States, Canada and Australia to justify not ratifying or even withdrawing from the legally binding agreement altogether.

Despite being signed in 1997, the Kyoto Protocol was only implemented in 2005 and ran until 2020, when it was superseded by the signing of the Paris Agreement at COP21 in 2015. After decades of diplomatic discussion and failures, a monumentally difficult agreement was reached that also included developing countries, with richer nations agreeing to provide US$100 billion each year in climate finance to help poorer nations adapt. It was the first time the world came together with the common goal of addressing climate change on a global level. Signed by 196 countries, the Paris Agreement was hailed by then US president Barack Obama at the time as being "the moment we finally decided to save the planet."

The scientific goals of the historic Paris Agreement are to keep global warming well below 2°C and as close as possible to 1.5°C above pre-industrial levels; for global emissions to peak as soon as possible; and to reach zero net global emissions in the second half of this century. All countries are obliged to commit to publicly disclosed mitigation targets from 2020 and to review targets every five years to ensure global reductions in greenhouse gas emissions. While all countries must contribute, developed countries with more resources at their disposal are expected to take the lead, ratcheting up ambition over time.

The Paris Agreement officially took effect on 4 November 2016, after at least fifty-five parties to the convention responsible for at least 55 percent of global emissions formally ratified the agreement. While it was an enormous political achievement, the Paris Agreement does not legally bind nations to achieve their emissions reduction plans, referred to as Nationally Determined Contributions (NDCs). The same non-binding status holds for specific commitments to provide financial support to developing countries to reduce their greenhouse gas emissions. It does not impose penalties like fees or embargoes for parties that violate its terms, and there is no international court or governing body that regulates compliance. In the absence of formal enforcement mechanisms, the effort relies on the political goodwill of nations to do the right thing. Ultimately, it operates through peer pressure and "naming and shaming" countries that are not working in the interest of the collective good. Many experts fear that without economic or political penalties for noncompliance, governments will never act fast enough to address the issue.

In November 2021, world leaders gathered in Glasgow, Scotland, to attend COP26, billed as the "best last chance" to avert planetary disaster and limit warming to 1.5°C. It was seen by many in the scientific community as the moment to set the course for the critical decade of transformation ahead. Delayed by a year because of the coronavirus pandemic, the COP26 meeting took place in the aftermath of the release of the first volume of the IPCC's *Sixth Assessment Report*. In a statement issued on the day of the IPCC report's release, the Secretary-General of the United Nations, António Guterres, delivered an urgent message to the world:

> Today's IPCC Working Group 1 report is a code red for humanity. The alarm bells are deafening, and the evidence is irrefutable: greenhouse gas emissions from fossil fuel burning and deforestation are choking our planet and putting billions of people at immediate risk. Global heating is affecting every region on Earth, with many of the changes becoming irreversible. The internationally agreed threshold of 1.5°C is perilously close ... The only way to prevent exceeding this threshold is by urgently stepping up our efforts and pursuing the most ambitious

path. We must act decisively now to keep 1.5°C alive. We are already at 1.2°C and rising. Warming has accelerated in recent decades. Every fraction of a degree counts. Greenhouse gas concentrations are at record levels. Extreme weather and climate disasters are increasing in frequency and intensity. That is why this year's United Nations climate conference in Glasgow [COP26] is so important ... If we combine forces now, we can avert climate catastrophe ... there is no time for delay and no room for excuses.

On cue, extreme weather ravaged the world during the Northern Hemisphere summer of 2021, vividly bringing our report to life. Europe sweltered through its hottest summer on record, with wildfires tearing through the Mediterranean; North America experienced unfathomable 50°C heat; rain rather than snow fell for the first time on record at the peak of the Greenland ice sheet; torrential rainfall washed away homes in China and western Europe, while severe drought crippled the southwestern United States, Madagascar and much of subtropical South America. In an overview of the IPCC report for the Australian newspaper *The Saturday Paper*, I wrote:

> Releasing our report against the backdrop of these disasters makes me hope people no longer need to use their imaginations to picture climate change – it's here, right now, part of life for every single person on the planet. The evening news now looks like a disaster movie. As a climate scientist, I know it's far worse than that – it's the reality of a rapidly destabilising world ... During my time working on the *Sixth Assessment Report*, it dawned on me that this IPCC assessment is the scientific community's last chance to really make a difference. If our latest report and the intensifying evidence now all around us doesn't convince this generation of political leaders that we must stabilize the Earth's climate immediately, nothing ever will. Let's hope the governments of the world are finally ready to listen.

Scientists had done our job; we presented governments with the best possible information needed to underpin their policies. Many of us worked well beyond our physical and mental limits through yet another wave of the COVID-19 pandemic as we scrambled to make the Final Government Draft deadline of midnight, 12 March 2021. After uploading the very final touches to version 92 of our chapter on water cycle changes at 11:48 p.m. Paris-time, I reflected on the achievement in my diary:

I've finally finished my IPCC edits. To say I am exhausted is an understatement. I can't remember having ever been so tired in my entire life. My head aches, my eyes are sore and I feel like I'm floating outside of my body. I couldn't imagine myself at this stage of the process, but here I am. It's hard to describe how deep I've had to dig to make it this far. Although I'm definitely not in good shape, I made it to the end. It's going to take a while to recover from the enormous pressure I've put myself under over the past three years. But going over the document again this morning, seeing the polished text that finally emerged from the rat's nest of markup has been a really rewarding experience. We have covered everything from trends in the west African monsoon, the melting of mountain glaciers in the Andes, the aridification of vast bands of the Earth – including southern Australia – it's hard to hold it all in your mind at once. Reading text that has been so carefully considered by scores of people multiple times has a real power, a real gravity. Every nuance has been considered and agonized over for thousands of hours by so many people. It really is a monumental feat. I'm proud to have been a part of such an extraordinary, truly international experience. Working with people from all over the world on the common goal of helping humanity grasp the seriousness and scale of the climatic destabilization that is happening has been an incredible experience – grueling, exacting, relentless – demanding everything

I have, stretching me well beyond the comfort zone of what I ever thought I could handle... While I wanted to do my best to contribute to this global process, it's taken a heavy toll on my body, my own research and my personal life. It's hard to believe this is almost behind me. I made it.

Our community moved heaven and earth to assemble a mountain of irrefutable evidence demonstrating the role of human activity on our climate and very clearly laid out all the risks each region of the world faces under a variety of possible futures. Our key message was: climate change is here, and we are causing it; but how bad we let things get is still up to us. Surely we had done enough to compel our leaders to act. The only thing left was for our political leaders to step in and save the day at Glasgow.

Given the travel restrictions associated with the coronavirus pandemic, like many, I did my best to follow the COP26 proceedings online through videos and social media feeds. I wept listening to the procession of stirring speeches delivered at the opening ceremony of the World Leaders Summit, knowing just how much was at stake. There was so much genuine conviction in every single person who took to the stage. Each word felt laden with profound historical significance; a moment that we may look back on as a critical make-or-break moment for humanity. Our beloved elder, Sir David Attenborough, channeled all the wisdom of his ninety-five years to passionately call on world leaders to seize the moment to restore life on our planet:

> The people alive now, the generation to come, will look at this confer-
> ence and consider one thing. Did that number [the concentration of
> carbon in our atmosphere] stop rising and start to drop, as a result
> of commitments made here? There's every reason to believe that the
> answer can be yes. If working apart we are a force powerful enough
> to destabilize our planet, surely, working together, we are powerful

enough to save it. In my lifetime, I've witnessed a terrible decline. In yours, you could and should witness a wonderful recovery. That desperate hope, ladies and gentlemen, delegates, excellencies, is why the world is looking to you and why you are here.

*   *   *

In the days that followed the inspirational opening ceremony of COP26, I held my breath as I watched the announcements emerging from the meeting. The good news was that seventy-six parties pledged net-zero emissions targets, mostly by 2050. This represents at least 75 percent of the world's total greenhouse gas emissions so it was a really significant step. Policy analysts calculated that if all the new net-zero pledges are fully implemented, then we are likely to limit warming to 1.4–2.8°C by the end of the century. While still not enough, it felt like a good start; a vast improvement on the 4°C of warming the world was facing in the lead-up to signing the Paris Agreement. There were also very encouraging signs of real progress, with fourteen regions, including the United Kingdom, the European Union, Canada and New Zealand, legislating their 2050 targets, making them legally enforceable.

Other countries, including Australia, the United States and Brazil, theoretically agreed to net-zero promises by 2050 in policy documents, but these pledges have not yet been backed up by legally binding commitments, and in some cases contain very questionable pathways to decarbonization. Significantly, big emitters like China, Russia and Saudi Arabia also agreed to net-zero emissions by 2060, and India by 2070. While these are all really positive steps in the right direction, the problem is that we are relying on the assumption that every nation will make good on their promises. This is a real issue for poorer nations that are already struggling with poverty; many of their plans rely on the US$100 billion per year in adaptation finance provided by developed nations to help them address climate change. The figure is actually quite conservative when you consider that the IPCC currently estimates that adaptation costs in developing countries alone could reach US$70 billion per year, with that figure climbing to US$140–300 billion in 2030 and US$280–500 billion by 2050. Despite wealthy nations agreeing to provide US$100 billion per year when the Paris Agreement was signed back in 2015, the level of funding required failed to materialize in Glasgow and has now been pushed out to 2025. This left many delegates from vulnerable nations bitterly disappointed, as many regions need urgent financial assistance to adapt to worsening climate change *right now*, not a decade after it was first promised in Paris. Despite UN experts saying the world must spend five to ten times more to help vulnerable people adapt to inevitable environmental upheaval, concrete commitments by rich nations to the climate finance fund are still shamefully inadequate.

Perhaps the most heavily criticized outcome of COP26 was not only that the current pledges are still insufficient, but the fact that they still include plans to burn fossil fuels well past what is possible to limit warming to 1.5°C above pre-industrial levels. The IPCC report clearly states that we know we need to reduce greenhouse gas emissions by at least 50 percent between now and 2030 to keep this goal within reach. The 2021 Global Carbon Budget report estimates that the world has around a decade of burning carbon at the current rate if humanity hopes to avoid catastrophic warming. Despite draft text in the Glasgow Climate Pact including a clause about "accelerating efforts towards the phase out of unabated coal power and inefficient fossil fuel subsidies," members failed to agree on a "phase out" of coal use, following last-minute objections by India and China that succeeded in watering down the language to a far weaker "phase down." Nonetheless, despite the compromised phrasing, it is important to note that it was the first time that a United Nations climate change declaration directly mentioned the need to rapidly reduce the use of fossil fuels, something that nations like Saudi Arabia and others have stripped out in the past. The international community has now very clearly signaled to financial markets that the fossil fuel era is coming to an end.

While acknowledging the critical diplomatic significance of the achievement, we are dealing with an emergency situation where actions must speak louder than words. Clearly there is still a real disconnect between the political response to climate change and the scientific reality we face. While it was very positive to see major powers like the United States and the United Kingdom bring significant commitments to reduce emissions this decade, other countries either did not increase their ambition at all, or only did so in a negligible way. For example, China, now the world's largest greenhouse gas emitter, responsible for around a third of global emissions, only agreed to increase its use of renewable energy to 25 percent by 2030, up from its previous pledge of 20 percent. While their Nationally Determined Contribution submission mentions the phase out of coal and their intention to stop building new coal-fired power projects overseas, it also stated "it is unlikely to fundamentally change the coal-dominated energy mix in the short term" and did not formally sign up to the clean energy transition statement. Further evidence that political leaders aren't quite ready to give up fossil fuels was also obvious from an effort to phase out oil and gas production, which only received limited support on the sidelines of the summit. Despite the dominant role of oil and gas in fueling climate change, the Beyond Oil and Gas Alliance garnered just twelve members, including Denmark, Costa Rica and France.

The IPCC report very clearly states that the large-scale burning of fossil fuels is the primary cause of climate change. Virtually all of the observed global warming we have experienced so far has been driven by emissions from human activities. According to the 2021 Global Carbon Budget report, between 1960 and 2020, around 82 percent of carbon dioxide emissions came from the burning of fossil fuels, with the remaining 18 percent generated from land use changes like deforestation and the degradation of ecosystems. Total anthropogenic emissions more than doubled over the last sixty years, with almost half of the carbon dioxide accumulated in the atmosphere emitted since 1990. The burning of coal was the largest contributor to these emissions, closely followed by oil and gas. As the latest IPCC assessment highlighted, greenhouse gas concentrations are now at their highest levels in at least 2 million years, leading to rates of warming unprecedented in thousands of years, with no sign of slowing down. Despite the Glasgow Climate Pact stating that it "expresses alarm and utmost concern that human activities have caused around 1.1°C of warming to date," the science is very clearly telling us we are still nowhere near doing enough to reduce greenhouse gas emissions. So what exactly is going on?

In a nutshell, there is still entrenched reluctance in many parts of the world to phase out coal, oil and gas. Without genuine commitments to eliminate the burning of fossil fuels – the factor responsible for the lion's share of the problem we face – we will lock in dangerous levels of climate change. Scientifically speaking, it is as simple as that. But when you consider things politically, things become much more complex. According to the International Monetary Fund, in 2020 governments spent US$450 billion in direct subsidies to the fossil fuel industry – four and a half times more than developed countries are willing to spend on financing climate change adaptation in vulnerable regions. Instead, nations like Australia, the United States and China are basing their policies on unproven, highly subsidized carbon capture and storage technology rather than investing in renewable energy solutions that exist today.

Most alarmingly, plans to continue to expand the fossil fuel industry are still underway – the United Kingdom currently has at least forty new projects in the pipeline, including plans to develop the controversial Cambo offshore oilfield in the North Sea. In the United States, President Joe Biden's administration's call for a pause on new oil and gas leases on federal land and water was blocked in the federal court after fourteen state governments, including Louisiana and Texas, legally challenged the move. As of November 2021, Australia had seventy-two new coal projects and forty-four new gas projects under development, including the exploration of the new Beetaloo and Canning gas basins in northern Australia, which are being heavily supported by government subsidies. These projects would more than double Australia's gas and coal production, resulting in 1.7 billion tonnes of greenhouse gases each year, equivalent to over 200 coal-fired power stations, or twice the emissions of global aviation. And this only includes projects currently under development, not prospective projects where no publicly available information exists to estimate the impact of their emissions. This colossal expansion of the fossil fuel industry is entirely inconsistent with the nation's stated target of achieving net-zero emissions by 2050.

While it is clear that the fossil fuel industry still heavily influences governments the world over, signs of resistance are emerging. One of the positive developments to come out of COP26 was an agreement to phase out coal-fired electricity in the 2030s for major economies and the 2040s for developing nations. While forty countries including the United Kingdom and Germany, signed the pledge, the world's biggest coal-dependent economies of the United States, China, India and Australia are missing from the deal. Similarly, another agreement to end all overseas finance for "unabated" fossil fuel projects – those run without carbon capture and storage – by the end of 2022 was only formally signed by thirty-one governments and financial institutions, including the United Kingdom, United States and Canada. Significantly, neither statement was signed by China, Japan, India or Australia, some of the world's largest fossil fuel emitters and exporters. This reflects the fact that our political leaders are still protecting corporate interests that intend to keep the fossil fuel industry on life support until the bitter end. Without genuine commitments to phase out the burning of fossil fuels, we are not serious about addressing climate change. The simple truth is that unless we consign coal, oil and gas to history, we will destabilize our climate and reconfigure life on the planet as we know it.

The most confronting aspect of COP26 was the overwhelming presence of fossil fuel lobbyists representing major oil and gas companies like Shell and British Petroleum. Over 500 people with links to the fossil fuel industry attended, contributing to at least twenty-seven country delegations including Australia, Canada and Russia. Together, they were the largest delegation present at the COP26 summit – more than any single country present at the meeting – far outnumbering the delegations from countries that are worst affected by climate change. According to Fijian prime minister Frank Bainimarama, Pacific Island negotiators were outnumbered by fossil fuel representatives by more than twelve to one.

To me as a climate scientist, the presence of hundreds of people openly promoting the continued expansion of the fossil fuel industry alongside people literally being displaced from their homes felt completely immoral. How can a climate summit pleading for world leaders to urgently avert planetary disaster be crawling with fossil fuel lobbyists? It's like the tobacco industry showing up to a lung cancer conference to promote their products. It's easy to feel demoralized by the fact that our political leadership is failing us. In the face of such insidious forces, is there anything any of us can ever really do to turn this around?

*   *   *

In the days following COP26, I thought long and hard about the leadership failure we witnessed at Glasgow. Like many people, I started to feel cynical and despairing about what unfolded. Hollow promises in the face of a global emergency; the cruelty of greed; the sting of knowing you've missed the moment. Ugandan youth climate activist Vanessa Nakate clearly summed up the implications of inaction, saying: "You cannot adapt to lost cultures, you cannot adapt to lost traditions, you cannot adapt to lost history, you cannot adapt to starvation. You cannot adapt to extinction." Addressing protesters gathered in Glasgow, she told the crowd: "Leaders rarely have the courage to lead. It takes citizens, people like you and me, to rise up and demand action. And when we do that in great enough numbers, our leaders will move."

And then something finally clicked. I realized that human history is an endless tug of war for social justice: a struggle between those wanting to maintain the status quo that protects the interests of a few, and others who fight for equality for all. This struggle has been part of every great victory in our history. From the abolition of slavery to the fight for women's rights, people have stood up against injustice and said enough is enough. Yes, our political leaders collectively failed us at Glasgow, but in the pavilions and out on the streets, leadership of another kind was also there thriving – our Indigenous leaders, innovators, young people, entrepreneurs, our environmentalists, scientists, artists, activists – visionaries of every stripe, rising to the biggest challenge humanity has ever faced, in any way that they could.

Despite the travel restrictions posed by a deadly pandemic, people gathered from every corner of the world to be part of this critical moment in human history. A colorful procession of 100,000 demonstrators filled the streets of Glasgow to demand more action inside the negotiation rooms. But this was just one part of the Global Day of Action for Climate Justice events that took place on 6 November 2021 in over 100 countries around the world, from Kenya to Canada, China to Brazil, Norway to Australia. And that's when it hit me – we are witnessing the biggest social movement of our time. A time of true global citizenry, driven by our passion to save the one thing that sustains us all: our Earth.

At the COP26 opening ceremony, Kenyan climate activist Elizabeth Wathuti powerfully reminded us of our shared humanity, asking world leaders to: "Please open your hearts. If you allow yourself to feel it, the heartbreak and the injustice is hard to bear." In times of crisis, we don't look into each other's eyes and see our differences; we recognize our shared humanity. We feel each other's pain, and do whatever we can to ease each other's suffering. We care about injustice and about righting the wrongs of the past. We understand that we need to be the type of person who strives to be compassionate, loving and kind. When our moral compass tells us that we have lost our way, we let our heart guide us back to what feels right and true. When we align our values with our actions, we are unstoppable. The force for good has the power to change the course of history; it always has, and it always will. Right now, we need to recognize that collectively we have the power to change our political leadership, wherever we find ourselves in the world. Every battle won is a step closer to winning the war.

Now, more than ever, we need to get behind people whose actions are genuinely aligned with their ethics. People who choose to create a brighter future, who refuse to buy into the dystopian inevitability that vested interests would have us believe is the only option. Our visionaries are already busy creating a sustainable future where the environment and social justice are restored. We just need to use our democratic power to help shape the course of our future. Because the truth is, there are good people everywhere. Even in politics. Some of them are doing everything they can to work within the imperfect systems we have in place, together as a global community. The scale of climate change is monumental; so are the changes needed to decarbonize every sector of the global economy and switch to clean technologies, especially in nations that have depended on exporting fossil fuels for decades or are struggling to lift their people out of poverty. System change is not something that will happen overnight; but its momentum is picking up speed, and soon gradual progress will become exponential. The social tipping point we need is now clearly on the horizon. It's precisely the moment when we can't give up. The darkest hour is always just before the dawn.

So while it is completely understandable to feel intense anger and frustration at the pace of change, we need to remember that although some leaders failed us at COP26, there were many, many who didn't. They showed up in the face of the enormous challenge and did everything they could. They were at the table, working around the clock to fight for a better world. The problem is, the people working within established systems are still outnumbered. In an era when it is so easy to be cynical and pass judgment on those in the room at COP26 (or involved in the IPCC process), it's helpful to consider the wisdom of former US president Theodore Roosevelt:

> It is not the critic who counts; not the man who points out how the strong man stumbles, or where the doer of deeds could have done them better. The credit belongs to the man who is actually in the arena, whose face is marred by dust and sweat and blood; who strives valiantly; who errs, who comes short again and again, because there is no effort without error and shortcoming; but who does actually strive to do the deeds; who knows great enthusiasms, the great devotions; who spends himself in a worthy cause; who at the best knows in the end the triumph of high achievement, and who at the worst, if he fails, at least fails while daring greatly, so that his place shall never be with those cold and timid souls who neither know victory nor defeat.

Unless there are more people willing to open their hearts and be moved enough to align their ethics with their actions, the status quo will prevail. In the words of Mahatma Gandhi:

We but mirror the world. All the tendencies present in the outer world are to be found in the world of our body. If we could change ourselves, the tendencies in the world would also change . . . We need not wait to see what others do.

Now is not the time to sit on the sidelines. It is the time to take a stand for what you really believe in. It is the moment to step up and do everything you can to influence the outcome of decisions that really matter. In democratic countries at least, our politicians are people we have voted for, people *we* have put in charge of *our* future. Each of us wields enormous political power in how we vote, where we put our money and what we say and do. Together, we can change governments, demand corporate responsibility and transform our communities one conversation at a time. We can create the change we want to see in the world.

10

A new kind of politics

IT'S EASY TO UNDERSTAND WHY people have become so disillusioned with politics. These days the pendulum seems to swing from one extreme to the next – from progressive globalism to regressive nationalism. In difficult times, people become increasingly tribal; we become suspicious of outsiders and our sense of solidarity and compassion for our fellow humans falters. This tendency influenced Britain's exit from the European Union and the resurgence of far-right groups like the Proud Boys in the United States and neo-Nazis in countries like Germany. When we view the world through such a divisive lens, we forget that it is possible to have a strong national identity while also contributing to the international community in constructive ways. Sometimes tuning into political debates can feel so "shouty" and demoralizing that it's hard to ignore the urge to switch off.

But the problem is, the more of us who disengage, the stronger these destabilizing forces become, as fewer people are there to rebalance the scales. Now more than ever before, we need to unite as a global community to restore peace and stability to our planet. Few issues highlight the urgency of the co-operative challenges we face more than climate change. Every nation in the world needs to work together to stabilize the Earth's climate. We either all succeed together or all fail together. You can think of addressing climate change as the mother of all group assignments that has been left to the last minute. The most challenging aspect is that the issue has long been politicized by vested interests seeking to protect the status quo of a fossil fuel–driven world. In many countries, corporate interests are closely tied to the funding of political parties, so they often have a disproportionate influence on decisions about how a nation's resources are used to generate economic wealth and social development. This process has proved itself, time and time again, to be open to interference by powerful corporations working together with corrupt government counterparts to push through developments that are not in the public interest.

As we saw at COP26, fossil fuel lobbyists are doing everything they can to convince people that they can continue to expand the coal, oil and gas industries despite what the science says. Pinning our hopes on non-existent or inadequate carbon capture and storage technology as the climate crisis escalates is sheer madness. It prolongs the burning of fossil fuels, which will contribute to overall cumulative emissions that will remain in the atmosphere for centuries after the industry has been shut down. As the latest IPCC assessment clearly shows, every tonne of carbon dioxide really matters. Planning to rely on speculative technology to save us is a reckless gamble that will undoubtedly saddle future generations with higher levels of warming that will be very difficult, if not impossible, to adapt to, particularly in low-lying regions of the world and areas already struggling with rising seas, water security issues and extreme heat.

Despite the mountain of scientific evidence that has now been compiled, trying to enact strong climate change policy in fossil fuel–dependent countries like Australia and the United States has proved to be political poison. In Australia – the world's largest exporter of coal and liquefied natural gas – there has been fierce opposition by the fossil fuel industry to phase out its activities, resulting in the downfall of five prime ministers over the past decade. It brought to an end the leadership of progressive leaders Kevin Rudd in 2010 and Julia Gillard in 2013. Similarly, Malcolm Turnbull's run as Liberal opposition leader ended in 2009 when he lost a leadership ballot centered on the government's Emissions Trading Scheme. Following the 2013 federal election, rabid climate change sceptic Tony Abbott took office as prime minister, before being replaced by the more progressive conservative Turnbull in 2015. As prime minister, Turnbull attempted to take a stand on strong climate policy, but in the end was replaced by the pro-fossil fuel conservative Scott Morrison, who as federal treasurer in 2017 infamously brought a lump of black coal into the Australian federal parliament, saying, "This is coal. Don't be afraid, don't be scared, it won't hurt you." He went on to strenuously ridicule those arguing for stronger climate policy, saying they had an "ideological, pathological fear of coal," rather than a genuine concern based on their acceptance of the indisputable scientific reality that fossil fuels are cooking our planet.

Under Morrison's leadership, there were plans to double Australia's fossil fuel production while actively undermining the expansion of the renewable energy sector, including redirecting finance to support clean energy projects through the Clean Energy Finance Corporation (CEFC) and the Australian Renewable Energy Agency (ARENA) into carbon capture and storage projects instead of wind and solar initiatives. This is on top of the staggering AUD$10.3 billion in government subsidies the fossil fuel industry received in 2020–2021, more than the entire budget of the Australian Army. This is also despite the fact that as the sunniest continent on Earth, Australia only generated 10 percent of the nation's electricity using solar power in 2020–2021. It is clear that a powerful group of fossil fuel lobbyists are shamelessly doing everything they can to protect their corporate interests in Australia, while continuing to actively block international efforts to address climate change.

In the United States, the debate around climate policy has been similarly toxic. After Barack Obama played a vital role in garnering international support for the signing of the Paris Agreement in 2015, a swing back to the far right resulted in climate change denier Donald Trump withdrawing from the global agreement on 2 June 2017. This saw the world's second-largest emitter of greenhouse gases responsible for around 14 percent of global emissions in 2020, and around a quarter of cumulative historical emissions since 1850, snubbing international diplomatic efforts to try to safeguard the planet's stability. The US withdrawal sent a very negative signal to other signatory nations that they also didn't need to honor their pledges. As one of the world's richest nations, the United States was also expected to provide a considerable portion of the US$100 billion in aid to developing nations to help them cut emissions and adapt to a rapidly changing climate. It was a very dark moment in the history of international climate diplomacy, one that caused many to despair and feel that all hope was lost.

But then something incredible happened. In the wake of Trump's decision to withdraw the US from the Paris Agreement, the United States Conference of Mayors quickly stepped in. They stated their strong opposition to Trump's action and vowed that American mayors around the country would continue their local efforts to reduce greenhouse gas emissions. At the time, the CEO of the organization, Tom Cochran, said: "The nation's mayors have never waited on Washington to act, and have been strong proponents of action on climate for decades. Mayors will continue to harness their collective power to continue to lead the nation on this critical issue, regardless of what happens at the national level." There are now 470 "Climate Mayors" – representing forty-eight states, 74 million Americans and some of the largest cities in the country, including New York, Los Angeles, Chicago and Houston – who have banded together to intensify efforts to uphold the goals enshrined in the Paris Agreement. It was extremely heartening to see that, even in the absence of national political leadership, local leaders were determined to surge ahead with the rest of the progressive world and support the global clean energy revolution. Within hours of being sworn in as the new president on 20 January 2021, Joe Biden signed an executive order to reinstate the US in the Paris Agreement, to the immense relief of the international community.

Like millions of people around the world, I wept watching President Biden and Vice President Harris sworn in. After an ugly election campaign that polarized the nation using fear-based nationalism, listening to a leadership team with such a hopeful vision not only for their country but also for the world, felt like sanity had finally been restored. It made me recognize that ultimately, no matter how difficult the struggle, human decency will eventually prevail. Poet Amanda Gorman offered an inspiring reflection on a deeply divided country in "The Hill We Climb":

And so we lift our gazes not
To what stands between us,
But what stands before us.
We close the divide,
Because we know to put
Our future first, we must first
Put our differences aside ...

Let the globe, if nothing else, say this is true:
That even as we grieved, we grew.
That even as we hurt, we hoped.
That even as we tired, we tried ...

For there is always light,
if only we're brave enough to see it.
If only we're brave enough to be it.

Biden's inauguration address also rose to the challenge, providing a moving reminder that the goal of unity must be at the core of everything we do at this historic moment in human history:

And together we shall write . . . a story of hope, not fear, of unity, not division. Of light, not darkness. A story of decency and dignity, love and healing, greatness and goodness. May this be the story that guides us. The story that inspires us and the story we tell the ages yet to come. That we answered the call of history. We met the moment . . . this is what we owe our forbearers, one another, and generations to follow.

What an extraordinary time to be alive; we are part of the generation that will help heal the world.

*   *   *

The struggle for justice, equality and peace is not new. Humanity's story is one punctuated by great uprisings, when people finally sense that enough is enough, forcing the status quo to change. One of the most iconic examples is the American civil rights movement, led by Dr. Martin Luther King Jr. In the face of violent racism towards African Americans by white supremacists, King led a nonviolent movement against racial segregation and discrimination in the southern United States during the 1950s and 1960s. The movement had its roots in the centuries-long efforts of enslaved Africans and their descendants to resist racial oppression and abolish the barbaric institution of slavery. Although African Americans were emancipated and granted basic civil rights following the American Civil War of the 1860s – fought between the southern and northern states over the abolition of slavery – the fight for equality was far from over. Through a series of nonviolent protests, African Americans continued to fight for their democratic right to vote and confront the injustice of ongoing racial oppression. On 28 August 1963, a century after President Abraham Lincoln signed the Emancipation Proclamation freeing the slaves, King climbed the marble steps of the Lincoln Memorial in Washington, D.C., to address over 200,000 people, of all races, who had gathered in solidarity to demand equal rights for his people. In his iconic "I Have a Dream" speech, he described his vision of America, reminding people of the urgency of using their democratic power to tip the scales of justice:

> This is not time to engage in the luxury of cooling off or to take the tranquilizing drug of gradualism. Now is the time to make real the promise of democracy. Now is the time to rise from the dark and desolate valley of segregation to the sunlit path of racial justice. Now is the time to lift our nation from the quicksand of racial injustice to the solid rock of brotherhood. Now is the time to make justice a reality for all of God's children.

In April 1967, one year before King was shot dead by a racist, he delivered his powerful anti-Vietnam War speech at Riverside Church in New York City, reminding people of the urgency of rising to the political challenges of their time. He spoke of the horrors of the war, saying, "This madness must cease," pleading for sanity to prevail. He went on to say:

> We are now faced with the fact that tomorrow is today. We are confronted with the fierce urgency of now. In this unfolding conundrum of life and history there is such a thing as being too late. We may cry out desperately for time to pause in her passage, but time is deaf to every plea, and rushes on. Over the bleached bones and jumbled residue of numerous civilizations are written the pathetic words: too late.

His prophetic words are as relevant for the struggles we face today as they were during the revolutionary 1960s. It makes me realize that at the heart of every transformational social movement is the desire for justice, unity and peace. My hope is that the social movement needed to protect all life on our planet will continue to strengthen and be inspired by the bravery and spirit of all those who have come before us. In her inspiring essay "Ten Ways to Confront the Climate Crisis without Losing Hope," American writer Rebecca Solnit summarizes the challenge we face:

> The world as we knew it is coming to an end, and it's up to us how it ends and what comes after. It's the end of the age of fossil fuel, but if the fossil-fuel corporations have their way the ending will be delayed as long as possible, with as much carbon burned as possible. If the rest of us prevail, we will radically reduce our use of those fuels by 2030, and almost entirely by 2050. We will meet climate change with real change, and defeat the fossil-fuel industry in the next nine years. If we succeed, those who come after will look back on the age of fossil fuel as an age of corruption and poison. The grandchildren of those who are young now will hear horror stories about how people once burned great mountains of poisonous stuff dug up from deep underground that made children sick and birds die and the air filthy and the planet heat up. We must remake the world, and we can remake it better. The COVID-19 pandemic is proof that if we take a crisis seriously, we can change how we live, almost overnight, dramatically, globally, digging up great piles of money from nowhere, like the $3 trillion the US initially threw at the pandemic.

My hope is that we rise to the greatest challenge of our time, as others have throughout history, and realize that we are the generation that will determine humanity's fate. As Martin Luther King Jr. reminds us, we are confronted with the "fierce urgency of now" that requires all of us to use our political power to stabilize the Earth's climate and reconnect with our shared humanity. Now is not the time to look away, thinking that what you do as an individual doesn't make a difference. Because the truth is, if millions of people all over the world get behind this, we will change the course of our history. It is time to transition away from the age of mindless consumerism and shameless disregard for the natural world, to an awareness that planetary destabilization is now at stake, threatening the very viability of generations to come. In *Hope in the Dark*, Rebecca Solnit writes:

> As citizens of the Earth, we have a responsibility to participate. As citizens massed together, we have the power to affect change, and it is only on that scale that enough change can happen. Individual choices can slowly scale up, or sometimes be catalysts, but we've run out of time for the slow. It is not the things we refrain from doing, but those things we do passionately, and together, that will count the most . . . Movements, campaigns, organisations, alliances and networks are how ordinary people become powerful – so powerful that you can see they inspire terror in elites, governments and corporations alike, who devote themselves to trying to stifle and undermine them. But these places are also where you meet dreamers, idealists, altruists – people who believe in living by principle. You meet people who are hopeful, or even more than hopeful: great movements often begin with people fighting for things that seem all but impossible at the outset, whether an end to slavery, votes for women or rights for LGTBQ+ people.

Now is not the time to start eulogizing our planet or to accept a foregone conclusion of an apocalyptic future. It is so important to remember that the future is not yet written – we are writing it right now. To give up at this moment would be to give in to the paralysis of cynicism and despair. We must resist the urge to curl up in the fetal position and declare that it's all too hard, or to wait for someone else to figure this out. We are not powerless, but we need to understand that the time for complacency has long passed. We are facing a planetary emergency; we need to act in the same way we would if the people we love were in grave danger. As Greta Thunberg said: "I want you to act as you would in a crisis. I want you to act as if our house is on fire. Because it is." The most influential thing we can do is use our political power to build the critical mass needed to restore sanity and care to our world to change the course of history.

We can't all be great leaders like Martin Luther King Jr., but we can show up and lend them our support when it really matters. Like those thousands of demonstrators who came to hear him speak at the Lincoln Memorial in 1963, we too can join the gatherings taking place in the cities and towns wherever we live, all over the world, and choose to be on the right side of history. As anyone who has been part of the contemporary Black Lives Matter protests for racial equality or the MeToo movement for women's rights will tell you, the fight for justice never ends. Veteran environmentalists will remember the fights to stop whaling, deforestation and the proliferation of nuclear weapons. Now we have climate change, an issue that transcends national boundaries and calls for humanity to come together as a global community. In this book I have tried to share everything I know about climate change, so that you are aware of the reality of the situation, so you can choose what you want to do next. I've tried to distill all the wisdom that has ever been passed on to me, in the hope that you will feel moved to do something – anything – to be part of the social tipping point we need to restore care and decency to this world. If you have the courage to connect your head with your heart, you will know what to do next.

*   *   *

There is a lot of truth in the saying "the standard you walk past is the standard you accept." In an era when people have become disillusioned with the entrenched inertia and corruption that seem to dominate politics, they are being left with no other option but to take matters into their own hands. One of the most exciting political developments of recent years, in Australia at least, has been the rise of independent candidates who are willing to take on issues that major parties are unwilling or unable to face. In a pro-fossil fuel country like Australia, trying to enact meaningful climate policy has led to a revolving door of political leadership. As the main parties have torn themselves apart trying to appease fringe factions, grassroots campaigns are getting behind sharp, ethically driven independent candidates attempting to restore balance, diversity and sanity to our parliament. They want an end to the culture of negativity, senseless division and elitism that has hijacked our political system. They believe that core community values should be better reflected in our parliament. They want common sense and a shared vision for our country reinstated. They understand that the choice is either to laze around and complain about it, or step up and fix it.

One of the most stunning political wins in modern Australian history was the overthrow of former prime minister Tony Abbott – who made Australia the only country in the world to *abolish* a national carbon pricing scheme when he seized power in 2013 – by independent candidate Zali Steggall, who based her campaign almost entirely on strong climate action. During the 2019 federal election, a number of local community groups channeled their disillusionment with business-as-usual politics into backing Steggall.

Steggall's campaign was in complete contrast to Abbott's climate change denial, speaking to many educated voters who did not need to be convinced about the threat of climate change in a country like Australia. Middle-class voters in Warringah, an affluent electorate in Sydney's northern suburbs, turned out in droves to support Steggall, leading to a landslide victory that saw Abbott – a man who has single-handedly done more to delay climate action in Australia than anyone in our political history – finally removed from a seat he held for twenty-five years. It was a profoundly symbolic victory that sent a very clear message to the rest of the country that everyday Australians are sick and tired of toxic politics. We want our leaders to do better. We want them to protect collective community values, not the corporate interests of the few.

The most inspiring aspect of Steggall's win was how strongly the local community got behind her. After starting with a few disaffected moderates, the grassroots campaign quickly swelled to form a local army of passionate supporters. They held intimate kitchen table meetings, inclusive community events and took to social media to convince their local community that Steggall was the real deal, and that voting for her was a powerful way to vent their deep concern about climate change and the dysfunctional downturn plaguing Australian politics. The community campaign resulted in genuine grassroots support for a progressive member of their own local community to win an "unwinnable seat" long held by Australia's conservative party. Instead of running multiple independent candidates, the community took the savvy unified approach of all rallying around Steggall to amplify their collective power.

Since then, Australia's independent grassroots political movement is fast gaining traction. It is a seismic shift in politics that is starting to return power to local people, where it has always belonged. More and more passionate people, notably smart and articulate women, are stepping up to give voice to their local communities. Their core values are centered around action on climate change, integrity to dismantle entrenched corruption, and gender equity for women and girls. People are motivated by wanting to change the culture of politics, which has been particularly toxic for women, Indigenous people, ethnic minorities and the LGBTQ+ community. Their aim is to restore a sense of ethics and basic decency to our democratic process. They believe that by having a diversity of people entering politics, it will no longer be the domain of wealthy and well-connected political forces seeking to protect the status quo, particularly on issues like climate change. Many hope it will usher in a more progressive, inclusive and pragmatic era of Australian politics.

At her campaign launch in November 2021, Zoe Daniel, independent candidate for Goldstein in Melbourne's wealthy inner south, clearly articulated her hope: "Imagine what a crossbench of community-minded independents could achieve from the sensible centre." When I attended one of her online "town hall" events in December 2021, the refreshing positivity and intelligence of both Daniel and her local community made me feel that the power of this political movement should not be underestimated. At a time when democracy is in crisis, we need our best and brightest leading with integrity. Our communities are hungry for change.

Although independents will never be able to outnumber members of the dominant political parties, they can sometimes hold the balance of power on critically important decisions. They also provide our parliament with a conscience. By demanding accountability and demonstrating moral integrity on the most vital ethical issues of our time, their strong stand may influence the more moderate members of the government or opposition to take a stance based on their personal values that might put them at odds with the rest of their party. Progressive alliances based on upholding our society's shared values – rather than simply toeing the traditional party line – become possible. Crossing the floor will no longer be taboo.

Only by finding common ground can we attempt to reduce the polarization that has seen the emergence of an aggressive, highly adversarial form of politics. When politics is polarized this way, quieter, more considered people from more diverse parts of our communities are discouraged from engaging in the political process, which further serves to entrench the status quo. Perhaps that's the point. But when there are more independent voices at the table speaking with clarity, intelligence and integrity about the things that really matter to the communities they represent, it forces others to reflect on their own behavior and how it aligns with their ethics. Clearly, our political system has lost the way; maybe, just maybe, the rise of the independents might help bring us back together by restoring our faith in each other and the democratic process.

It is clear that Australian voters are hungry for this long-overdue shift in our political system. Around fifty independent groups formed the "Voices Of" movement to contest the 2022 Australian federal election. A group called Climate 200, spearheaded by Australian renewable energy expert Simon Holmes à Court, was established to provide funding, legal, strategic and political support to independent candidates "who stand for cleaning up politics and following the science on climate change." They crowd-funded around AUD$12 million to help level the political playing field for grassroots independents.

In a defining moment in Australia's political history, Morrison's right-wing government was emphatically voted out, and an extraordinary twelve parliamentary seats in the House of Representatives were won by independent candidates in the 2022 federal Australian election. The phenomenal success of the grassroots movement took the major political parties by surprise as previously "safe" conservative seats were swept away by an unprecedented groundswell of community support for independents right around the country. The Australian Greens Party also recorded their strongest electoral result ever, claiming four seats in the lower house and twelve in the senate, strengthening the crossbench support needed for real action on climate change. The long-overdue "climate election" that many disillusioned Australians had been waiting for finally arrived. This collective triumph has ushered in a new era of progressive politics in Australia, one that may – at long last – see the end of the political deadlock that has plagued the nation's climate policy for close to a decade.

Given Australia's historical pro–fossil fuel stance, no doubt the world will be watching to see if our nation can finally join the rest of the progressive world and enact credible climate policy. Unfortunately, the policies of both sides of politics still support the continuation of coal mining, the expansion of the natural gas industry, and ongoing government subsidies for fossil fuel use in Australia. This is why the increased presence of independents and Green party members in parliament matter more now than ever before. These new independents have formed an alliance that will help transform their communities' deep dissatisfaction with the business-as-usual model of politics into genuine social change. Many of these grassroots groups have regular town hall meetings to enable public conversations about politics, and social events to strengthen a local sense of camaraderie and belonging. Although the road ahead will no doubt be challenging, it's the emergence and strengthening of this social movement to restore decency to democracy that really matter.

Clearly, it will take more than one election in one country to turn the tide on this. But witnessing this new wave of politics in fossil fuel–obsessed Australia gives me hope that when we come together on issues we really care about, we genuinely have the power to influence our political systems. We can restore integrity and hope in the democratic process, inspiring people who have never been politically active before to feel like they can actually make a difference. Because at the end of the day, as former American president John F. Kennedy reminded us: "If not us, who? If not now, when?" The grassroots political movement springing up not only in Australia, but all over the world, gives me hope that if enough people care, we might be able to change the world after all.

11

Life imitating art

I HAVE TO CONFESS THAT when I started writing this book, I honestly didn't know how I was going to find my own way through. After years of grueling work on the IPCC assessment report, witnessing escalating extremes savage my country, and year after year of political failure, all I could see was a horrifying, bleak landscape stretching out before me. After compiling the material on ecosystem collapse, climate refugees and intergenerational damage, eventually I succumbed to a serious bout of depression. When I find myself weeping every day at the smallest things, I know it's time to step back and regain my equilibrium. As I'm slowly learning how to take better care of myself, I knew what I had to do: in the words of American writer Cheryl Strayed, I needed to "put myself in the way of beauty."

I swam in the cool ocean each morning, faced my inner demons on my yoga mat, and spent time with my girlfriend who held me as I cried. From my studio, I watched blue-masked honeyeaters feast on blood-red palm berries, rain soak the tropical jumble of our garden, and purple sunsets bruise the saturated spring sky. I ran along isolated beaches until my lungs felt like they'd catch fire. At work, I returned to the scientific research I had put aside to write this book, distracting myself in familiar routines I've used throughout my professional life to help find meaning and purpose in my days. As someone immersed in the science of climate change, it's easy to lose sight of the beauty in life and what the good people of the world are doing to try to turn this around. So on the weekends, instead of trying to digest the latest science, I started reading about history, politics, art, culture, psychology and spirituality. I read poetry and fiction that penetrated my numbness, gently reawakening my senses. I listened to my favorite pianist up loud as fierce thunderstorms cracked and flashed throughout the night. I made love with my golden-tanned man in the rain. I did these things until I felt still and whole once more.

And then one afternoon, after weeks reconstituting myself with books, music, films and poetry, something finally dawned on me. A thought crystallized with the clarity of a clear winter's sky – everything we need to fuel the social movement already exists. We are creative, sensitive, imaginative humans who can use art to inspire a shift in the emotional world not only of ourselves, but of others. Creative expression has always been the bridge between the head and the heart. And the best part was realizing that the cultural evolution we need to change the world is already spreading through literature, galleries, cinemas, theaters, festivals and community halls, right across the planet. This realization felt like stumbling across the missing link; something that once found, connected all the dots. When I could finally see that the cultural awakening we need is happening right now, everything clicked. I could finally see my way through the forest again. It felt so fundamentally true that it pierced through the soul-deep layers of exhaustion, overwhelm and crushing despair that I'd been experiencing for years. It created a visceral sense of relief in my body, like an impossibly heavy burden being lifted from my shoulders. It was the arrival of the cavalry on the horizon, one that I thought had long forsaken us. For the first time in a long time, I wept tears of joy.

Sometimes people complain that "nobody is doing anything about this," but it's simply not true; they just aren't looking in the right places.

A critical mass is forming and is starting to gain real momentum. Despite what we might see on the news or read on our social media feeds, the truth is really this: over recent years, the climate movement has grown in power, sophistication and inclusiveness and has won many important battles. Maybe we just haven't been very good at telling these stories because we are too busy turning our attention to the next active front. Or maybe we have simply failed to imagine a better future because the conversation has been dominated by economics, energy policy and divisive politics, rather than seeing things through the far more human eyes of the shared values that form our cultures.

We have forgotten that our history is full of powerful social movements – driven by people motivated to reinstate a sense of care and basic human decency – that have changed our world, time and time again. We have centuries of cultural evolution to tap into; empathy has always been the antidote for a desensitized world. The way back to our heart, and to each other, is through art. Reconnecting with this elemental truth was a revelation. The "how" of the social movement finally made sense: we need to rehumanize our world. To do this we need creatives and storytellers of all kinds. Since that inspired afternoon, apparently I've become chirpier and have started singing around the house again. I catch my husband looking at me with the expression of watching someone you love finally come back up for air: a mix of relief, concern and sorrow.

But when you are confronted by the alternative reality of greed, ignorance and hatred, it's easy to understand how quickly you can become saturated with hopelessness and a sense of disillusionment with the human race. Instead of wielding our immense collective power for good, it's demoralizing when you look around and see that apathy and selfishness still prevail. In the words of British poet Zena Edwards, when we don't bother to "lift our face from the trough of consumerism" and reflect on our collective moral predicament, we fail to have the emotional response we need to face the planetary crisis we are now in. We need to reassess our shared cultural values to redefine what it means to be human at this moment in time.

In a wealthy country like Australia, sometimes it's easy to feel judgmental and impatient with the inane things people fixate on; it seems like such a frivolous waste of the freedom and privileges many only ever dream of. The evening news shows the growing refugee crisis spreading around the world, from southern Europe to east Africa, the Middle East to Latin America. Children sleeping in cardboard boxes in makeshift camps with no sanitation on the borders of safe havens, their parents pleading for anyone to show them some mercy, to help them restore their dignity. Many are fleeing endless conflicts that have crippled the economic and social development of some regions for decades. Their kids struggle to stay alive and healthy, let alone get an education or dream about anything beyond having a safe place to sleep. As a daughter of Egyptian migrants, the specter of poverty is seared into my DNA. It is part of the subconscious forces that motivate me to keep going every day: I do something because I can. If you are holding this book in your hands, you are probably in a better position than most to do something too.

Despite the blessings of a stable government and economic prosperity, many people in Australia are very politically disengaged: we are collectively more interested in the latest sport results and celebrity chefs than we are about the death of the Great Barrier Reef or the aridification of our fragile continent. In a global study of twenty-two countries, Australians were identified as the heaviest drinkers in the world, spending more time drunk in 2020 than any other nation. Australians drank to the point of being drunk an average of twenty-seven times a year, almost double the global average. We've seen an increase in the involvement of alcohol with family violence callouts. Maybe it was a way of coping with the stress of the pandemic, or maybe we are a hedonistic nation that gets wasted and lashes out when things get too hard. Whatever the case might be, the way we treat each other mirrors the way we treat the rest of the natural world.

We perpetually distract ourselves with trivial things that temporarily improve the comfort of our personal lives, keeping us from deeper self-reflection and meaningful discussion about the shared values we have as a society. In a world that has become increasingly addicted to screens and disconnected from nature, we have lost touch with the Earth and now tread heavily on it. British psychologist Sally Weintrobe has written extensively about the psychology of climate change, saying that we have normalized a "culture of uncare" that has accelerated during the rapid period of globalization we've experienced since the late 1970s. She argues that our culture of mindless consumerism and entitlement is driven by a selfish belief that the Earth is here "solely to provide endlessly for us and to absorb all our waste." Weintrobe explains how trillions of dollars have been spent on undermining our intrinsic human capacity to care for the environment through political framing and manipulation by the mass media and advertising. This has led to a dangerous distrust of science that has delayed an appropriate, pragmatic response to climate change, instead leaving us scrambling to face an escalating emergency without the psychological equipment we need to face it.

This culture of uncare has also led to us becoming a more materialistic and narcissistic throwaway society that values fulfilling an individual's immediate desires over safeguarding our collective future. During the 1960s, economist Garrett Hardin termed this the "tragedy of the commons," where self-interest drives individuals to exploit collective resources in the short term, even to their long-term detriment. In his seminal 1968 paper published in the journal *Science*, Hardin wrote that caring for humanity's shared future required a "fundamental extension in morality." Instead, we seem to be endlessly distracting ourselves with inane things that divert people's attention from using their individual power to drive the social and political change we need to restore balance and sanity to our world. Instead of being active citizens, many people are choosing to be passive consumers in the face of the global crisis.

In his scathing article "Capitalism Is Killing the Planet," George Monbiot describes a media study that found that the word "cake" was mentioned ten times more often than "climate change" on UK television in 2020. The word "dog" was mentioned 170 times more often than "biodiversity." Despite being the most critical technology that will decarbonize the global economy, the terms "wind power" and "solar power" combined were heard less frequently than "banana bread." Monbiot goes on to say:

> The current ratio reflects a determined commitment to irrelevance in the face of global catastrophe. Tune in to almost any radio station, at any time, and you can hear the frenetic distraction at work. While around the world wildfires rage, floods sweep cars from the streets and crops shrivel, you will hear a debate about whether to sit down or stand up while pulling on your socks, or a discussion about charcuterie boards for dogs. I'm not making up these examples: I stumbled across them while flicking between channels on days of climate disaster. If an asteroid were heading towards Earth, and we turned on the radio, we'd probably hear: "So the hot topic today is – what's the funniest thing that's ever happened to you while eating a kebab?" This is the way the world ends, not with a bang but with banter.

While I am the first to agree that we all need a break from facing difficult realities sometimes, if we can't dip below superficialities and deeply engage with the more confronting truths that lie beneath, it will be hard to reach the social tipping point we need to secure the survival of our species. Monbiot argues that instead of exercising real political power in the face of the global catastrophe, people are instead preoccupied with what he calls "micro consumerist bollocks" – the smaller issues of plastic straws, shopping bags, and coffee cups, instead of turning our attention to the infinitely more important social and political forces that are threatening to destroy our life-support system. He also highlights the hypocrisy of people's consumer behavior in light of the scientific reality of the crisis we face:

> We are rightly horrified by the image of a seahorse with its tail wrapped around a cotton bud, but apparently unconcerned about the elimination of entire marine ecosystems by the fishing industry. We tut and shake our heads, and keep eating our way through the life of the sea.

Clearly there is a disconnect between what ordinary people feel they can do in the face of the climate crisis and what needs to be done. While lifestyle changes are certainly important, they are far outweighed by the industrial-scale need for detailed, legally binding policies to urgently phase out the burning of fossil fuels and rapidly restore the Earth's ecosystems. Fossil fuel interests have sought to deflect the conversation towards personal responsibility: the car you drive, what you eat and the lifestyle you lead. While all of these things are undoubtedly important – these lifestyle choices make us healthier, save us money and make us feel better about our impact on the planet – ultimately it makes people feel like climate change is their fault, rather than keeping the pressure on governments to enact policies that hold corporate polluters accountable and pass legislation that will lead us towards decarbonizing our economic systems.

Shifting the blame from the criminal actions of destructive industries to individual behavior has had the insidious effect of many people feeling powerless to do anything meaningful to address climate change. But when we fall for this trap, we allow "business as usual" to continue unopposed by the very people who ultimately create or remove the social license needed to keep trashing our planet. When we wake up to the fact that we collectively have the power to transform our social and political systems, we can correct our course and steer humanity out of the treacherous seas of greed and apathy, back to the safe shores of sanity and wisdom. We just need the moral compass of our shared humanity to guide us home.

*   *   *

The media is responsible for shaping our public conversations about climate change, creating the cultural conditions that support or undermine the sense of urgency needed to address the problem. Despite the fact that there are very high levels of concern about climate change in countless local communities, the mainstream media in many countries around the world are missing the opportunity to tell stories that will help educate and inspire people to be part of the transformative decade that lies ahead. This has left public conversations paralyzed by highly politicized, pessimistic discussions about economic loss rather than the sense of optimism and opportunity that the transition to an environmentally sustainable world will bring. In a provocative 2019 article by American journalists Mark Hertsgaard and Kyle Pope, "The Media Are Complacent While the World Burns," they call out the tendency of media professionals to avoid covering climate change because it is a "ratings killer":

> Yet at a time when civilization is accelerating toward disaster, climate silence continues to reign across the bulk of the US news media. Especially on television, where most Americans still get their news, the brutal demands of ratings and money work against adequate coverage of the biggest story of our time. Many newspapers, too, are failing the climate test. Last October [2018], the scientists of the United Nations' Intergovernmental Panel on Climate Change (IPCC)

released a landmark report, warning that humanity had a mere 12 years to radically slash greenhouse-gas emissions or face a calamitous future in which hundreds of millions of people worldwide would go hungry or homeless or worse. Only 22 of the 50 biggest newspapers in the United States covered that report.

They go on to say that you can't solve a problem by ignoring it. The role of journalism is to help the public grasp the seriousness of the issue, which should no longer be swayed by polarized politics. They write:

> There is a runaway train racing toward us, and its name is climate change. That is not alarmism; it is scientific fact. We as a civilization urgently need to slow that train down and help as many people off the tracks as possible. It's an enormous challenge, and if we don't get it right, nothing else will matter. The US mainstream news media, unlike major news outlets in Europe and independent media in the US, have played a big part in getting it wrong for many years ... This journalistic failure has given rise to a calamitous public ignorance, which in turn has enabled politicians and corporations to avoid action.

Bill McKibben, the American environmentalist who published the first mass-market book on climate change, *The End of Nature*, in 1989, has said that, journalistically speaking, the topic is "an exciting story filled with drama and conflict. It's what journalism was made for." The David and Goliath appeal of the underdog succeeding is inspiring for most people, as it highlights that – even in the face of enormous challenges – sometimes justice prevails. It reminds us that we are not powerless. There are the battles that are being fought all over the world right now, yet it is a fight that has only received sporadic, simplistic coverage by the mainstream media. This failure of news organizations to adequately cover the story has resulted in a lack of information available to the voting public. Although there is some outstanding reporting on climate change, namely by outlets like *The Guardian, The Conversation, The New York Times* and *The Washington Post,* much of this excellent in-depth coverage never reaches mainstream consciousness.

In fossil fuel-obsessed Australia, media tycoon Rupert Murdoch's News Corporation outlets, including *The Australian* newspaper and Sky News television, continue to give platforms to outrageous climate change sceptics who have profoundly shaped the public debate, resulting in ruinous political inaction. Factual news reporting has been increasingly conflated with editorial opinion, resulting in serious misinformation. This destructive influence has undermined the public's understanding of the issue, which in turn, has influenced politics. In April 2021, former prime minister Malcom Turnbull went as far as to say that the Murdoch media is "the most powerful political actor in Australia" - surpassing the influence of the country's two main political parties - which he says now poses a real threat to Australian democracy. Although there was a shift in News Corp's editorial policy in 2021 towards calling for action on climate change, many people remain suspicious. But perhaps the shift reflects the fact that the moral justification for Australia's position as the last stronghold of fossil fuels in the developed world is fast eroding. As the lived experience of worsening wildfires, drought, floods and ecosystem collapse continues to assault the nation, perhaps News Corp is simply playing catch-up with the high level of concern about climate change in our communities.

The Murdoch-owned media have been similarly destructive in the United States, where serious damage has been done to democracy through relentless climate change denialism and other far-right political extremism aired by outlets like Fox News. As far back as 1991, Western Fuels Association launched a series of misinformation campaigns to "reposition global warming as theory (not fact)" in the public's understanding of the issue. In a memo from communications strategist Frank Luntz leaked in 2002, he advized Republicans "to continue to make the lack of scientific certainty a primary issue in the debate." Luntz went on to say that "Should the public come to believe that the scientific issues are settled, their views about global warming will change accordingly." Instead of treating the topic of climate change as the scientific reality that it is, the Murdoch media have weaponized it as a political issue, often pitting genuine climate scientists against pro–fossil fuel spokespeople or morally challenged fake experts with no credentials in the field. By providing equal attention to non-evidence-based opinion littered with cherry-picked factoids, logical fallacies and conspiracy theories, these outlets have caused understandable public confusion about the true role of human activity in global warming. Under the guise of journalistic balance, this false equivalence has resulted in heated public debates that have served to undermine public under-standing of the issue, contributing to the deeply divided opinions on climate change that we still see today.

As American climate scientist Michael Mann explains in *The New Climate War*, now that the outright denial of climate change simply isn't credible anymore, instead there has been a concerted campaign of misinformation that continues to skew the public's understanding. He details how the forces of deception, delay and distraction have sought to shift the responsibility from fossil fuel corporations to individual action. The blunt truth is that 63 percent of cumulative worldwide emissions of industrial carbon dioxide and methane released since the start of the Industrial Revolution can be traced back to ninety major fossil fuel companies, known as "carbon majors." Since 1965, the top twenty emitters alone are responsible for releasing 35 percent of energy-related carbon dioxide and methane emissions. This includes investor-owned corporations like Chevron, ExxonMobil, BP and Shell, and state-owned companies including Saudi Aramco and Russia's Gazprom. As the study's author, American energy analyst Richard Heede, says: "These companies and their products are substantially responsible for the climate emergency, have collectively delayed national and global action for decades, and can no longer hide behind the smokescreen that consumers are the responsible parties." Allowing this to happen has been a key moral failure of our political systems but is one that is starting to be increasingly challenged in landmark legal cases against fossil fuel companies in places including the Netherlands, the United States and Australia. These legal cases are a step in the right direction that will help "keep the bastards honest," as we so eloquently say here in Australia.

Mann is also scathing about the exaggeration of the threats posed by climate change by so-called "doomists," who claim that the situation is already so bad that nothing can be done. Although it is often the work of ill-informed social media commentators or journalists with no qualifications in climate science, sometimes doomism even arises within the academic research community. The idea that the climate change problem is fundamentally unsolvable is not only scientifically wrong and unhelpful, but it tragically leads people down the same path of disengagement and inaction that maintains the social license for fossil fuel burning to continue. He calls this new form of denialism the "new climate war"– one that the planet is currently losing. That's why Mann and other prominent North American climate scientists like Katharine Hayhoe and Peter Kalmus feel compelled to speak out and counter misinformation by directly providing the public with the truth about climate change through books, podcasts and public talks in the limited spare time that they have outside of conducting their research. They do this as an additional service to the scientific community – even if it makes them a target for ongoing harassment and ridicule – because they know what is at stake.

When trying to grasp the true extent of the climate crisis, the most useful thing you can do is to educate yourself with the best available science – ideally straight from the horse's mouth – to help counter the misinformation that exists in your personal networks and online communities. If left unchecked, weeds of misinformation will choke the growth of the grassroots movement needed to mobilize people to act. The same way the vast majority of us would call out racism or misogyny, we need to become someone who can't walk past the lies and misinformation that continue to delay our collective response to the climate crisis. We can choose to be part of a tireless force for change that fights for truth, sanity and basic human decency to prevail.

*   *   *

Although I work as a scientist, it's my love of the arts that keeps me afloat. Virtually all of my friends are creatives of some kind: my inner circle is blessed with musicians, writers, poets, visual artists and dramatic performers (plus one divine yoga teacher). That's why creatives got a special mention in the introduction! My husband and I are both musicians; it's how we met more than twenty years ago, over an impromptu jam in a Sydney warehouse where my best friend – the guitarist in our all-girl rock band in high school – lived at the time. He still composes and records, but as you might imagine, time has been tight for me in recent years. But when I travel for work, I often take refuge in local art galleries to absorb their tranquility and allow myself to be transported into the world beyond words. I also love the emotional power of cinema to reconfigure my perspective. But as confessed earlier, I am definitely a word nerd – writing is the thing I do to balance my inner and outer worlds. It's the only thing I can still manage when all the color has drained out of my life.

So if you are a creative and have made it this far, I want to thank you for hanging in here with me. Really, thank you! As a sensitive person, I know that confronting technical material is sometimes hard to bear, but I hope you can now see why this social movement needs you. Your ability to feel things is your superpower, and that is exactly what the world needs right now. Unless we experience things on an emotional level, it is hard for people to care about a topic like climate change that can sometimes feel so huge and overwhelming. I know that no matter how many facts and figures I give people, in the end it is probably going to be a book, an artwork, a song, a photograph, a play, a performance or a film that eventually helps reawaken their sense of care for other people and the natural world. Art has always been the most powerful portal into the world of our emotions. It helps us imagine a world we cannot see. It gives us the images and language we need to ignite the emotional connection that will fuel the personal action we all need to take to turn the tide on this global problem.

In his essay "What the Warming World Needs Now Is Art, Sweet Art," Bill McKibben highlights the crucial role of creatives at this particular moment in time:

> If the scientists are right, we're living through the biggest thing that's happened since human civilization emerged. One species, ours, has by itself in the course of a couple of generations managed to power-fully raise the temperature of an entire planet, to knock its most basic systems out of kilter. But oddly, though we know about it, we don't know about it. It hasn't registered in our gut; it isn't part of our culture. Where are the books? The poems? The plays? The goddamn operas? Compare it to, say, the horror of AIDS in the last two decades, which has produced a staggering outpouring of art that, in turn, has had real political effect. I mean, when people someday look back on our moment, the single most significant item will doubtless be the sudden spiking temperature. But they'll have a hell of a time figuring out what it meant to us.

While the creative industries have risen to the challenge since McKibben wrote his provocative piece in 2005, clearly we still haven't reached the social tipping point needed to generate the global political response to address the climate crisis. Many people are not inspired by the ins and outs of economic and energy policy, so we need art to help us make an emotional connection with the issue.

That's where you come in. How do we emotionally and culturally make sense of the changes we are living through? Art has always been humanity's best way of processing the world around us. Artists have always used their sensitivity and imagination to help people feel something that compels them to act. Because the truth is that we only respond to the things we care about; art is the most powerful way of making the invisible visible. We need artists to help scientists translate cold, hard facts into raw human emotions. When an artist is generous enough to share their private processing of the world through their work, it helps people tap into something universal. Art helps break down the huge, intangible issue of climate change into a more manageable human scale. Although we all experience the same human emotions of joy, fear, sadness, disgust and anger, we all process things in our own way. Sometimes, sharing your personal response can have a powerful impact on others; it can help people articulate the complexity of emotions they also feel, but may not be able to express. That's what I hope to do with my writing. We need more individual artists and collectives to help drive the social movement we need to turn this around.

The good news is that some of this incredibly inspiring work is already well underway. I've been lucky enough to collaborate with the Melbourne-based initiative CLIMARTE, whose mission is to "harness the creative power of the arts to inform, engage and inspire action on climate change." Since 2010, they have sought to use art as a nonthreatening way for people to consider their personal response to our changing world. They recognize that art is a way of creating empathy, emotional engagement and the cultural understanding that is needed to bring about effective political change. To help bring this to life, they ran the ART + CLIMATE = CHANGE festivals in 2015, 2017 and 2019, gathering artists, galleries, museums and scientists to develop provocative exhibitions for the public. Following a hiatus caused by coronavirus lockdowns, in November 2021 CLIMARTE opened the world's first Climate Emergency Gallery to present groundbreaking participatory exhibitions and a public-facing events program to strengthen the ties between art, science and civil society. It's a very exciting development in the art world that I hope goes a long way to inspire the social change we need.

In the United Kingdom, Julie's Bicycle, an initiative established by the music industry in 2007, was founded to inspire those working across the arts and cultural sectors to mobilize people to act on the environmental crisis. They believe that art and culture must be at the heart of how we address climate change. Although we need science to underpin our political response, we also need to unlock the potential of the creative industries to inspire people to bring about the societal change that will fuel political change.

Julie's Bicycle's aim is to use the arts to inspire audiences to help create new ways of living in a rapidly changing world. They use the collective strength of the cultural community to try and unify artists into a social movement they refer to as the Creative Climate Movement. The initiative includes the Tate Galleries, Universal Music, Warner Music and the British Film Institute. As well as artwork and activism, Julie's Bicycle also developed the Creative Green Tools calculator to help arts and cultural organizations in over fifty countries to understand the environmental impacts of their venues, offices, tours, productions, events and festivals, allowing them to improve the sustainability of their operations. The initiative shows how the entirety of the arts sector – from artists themselves through to management – is showing up. It's an inspiring example of how the best kind of leadership always springs from our grassroots communities.

Another positive trend is that art is becoming an increasingly important part of UN climate summits. For COP21 held in Paris in 2015, a collective called ARTCOP21 brought together 550 events over fifty-four countries to help reframe climate change as not just a scientific or policy issue, but a cultural one. Through a series of installations, plays, exhibitions, concerts, performances, talks, conferences, workshops, screenings, gatherings and demonstrations, creatives banded together to highlight the need for strong climate action. They understand that a cultural revolution is needed to drive political change.

At COP26 in Glasgow in November 2021, artists gathered in person and online for the Climate Fringe festival, which was run "by civil society for civil society" and included exhibitions, lectures, public artworks, film screenings and poetry readings. Provocative billboard campaigns were run across the UK and Europe to target the greenwashing propaganda of corporations promoting questionable carbon offsetting schemes and the climate policies of corporations like Shell, NatWest and Barclays. Outside the negotiation halls, street artists installed satirical political posters in bus shelters and painted provocative murals like Bobby McNamara's *There's No Place Like Home*. Meanwhile, Extinction Rebellion's Red Rebel Brigade, a group of performance artists who shroud themselves in blood-red cloaks to symbolize the common blood that we share with all species, mesmerized onlookers as they strode silently through the streets.

Confronting artistic work about climate change is now popping up all over the world: in established galleries, street art, regional theaters, off-Broadway plays, concert halls, literary fiction and poetry. It's another way for people to engage on a level that isn't possible using scientific facts alone. Art has always been a powerful catalyst for social change. We need creatives of all stripes to step up and help guide us through this transformative cultural moment. My hope is that the sensitivity and vision of our creative communities will ignite our imaginations and stir our hearts enough to further fuel the social movement already sweeping our world.

*   *   *

We will not see the political response we need to address climate change until we redefine the cultural and social norms that are destroying life on Earth. Individual voters are responsible for creating or removing the social license needed to maintain the status quo of burning fossil fuels to the point of planetary instability. As we start to see democracy being undermined around the world, we don't have the luxury of being apolitical. We are at a critical crossroads, where it is important that we don't become disengaged from political processes that will determine our future. We need to do what we can, in our own way, using our own voice, to meet the crisis we face. We are living in an age of anxiety, exacerbated by the compounding threats of an uncontained pandemic and accelerating climate change. Everyone is dealing with feelings of fear, frustration and anger. It's important we find ways to talk about this collectively, as a community. The last thing we want to do is fall into a state of apathy or to become numb to each other's pain and struggle. It will be a far darker world if we lose our ability to find empathy and compassion. This is why we need artists, writers, poets and filmmakers: they can dismantle the walls of numbness that are being erected against the world as we start to silo off from our fellow humans and the rest of the planet.

In the absence of a collective religion, many people use art as a universal source of reflection, ceremony, ritual and shared meaning. If we can see each other and ourselves as entwined with the rest of the living world, it will help us value it in a way that is needed to shift our behavior. The beauty of art is that it doesn't belong to any particular group, country, gender or race – art belongs to everyone. We live in an age awash with more information than we can possibly process; as a result, we have little knowledge and even less wisdom. We have forgotten that only when we connect our head with our heart will we find true wisdom. As a scientist, I know it is easier to explain how bad a situation is, but far harder to actually feel it. That's where art and storytelling come in. They allow us to have an emotional connection with other humans and the rest of the natural world. They help us feel something, which might trigger a personal experience that leads to broader cultural change in our communities.

We need to have important conversations about what we will and won't tolerate as a society. This is a moment for us to redefine our shared values and place them at the heart of our decision making. Just because economic models don't include the value of trees, ecosystems, the ocean, or the entire biosphere, doesn't mean these things are worthless. Many of us believe that the value of the web of life that sustains us is immeasurable: you can't put a price on the life force that animates our world. We aren't just consumers and workers feeding endless economic growth; we are human beings. Art helps us re-evaluate what we value as a society and how far we have strayed from our inherent humanity.

Our cultural systems and the creative industries have huge capacity to shift public opinion and be a catalyst for real and lasting social change. Right now, all over the world, women, Indigenous communities and young people are stepping up and finally having their stories heard. We need to listen to our hearts and the hearts of others. Otherwise, we will end up living in a world full of greedy, soulless people. Because ultimately, the climate crisis is not just about percentages and parts per million. It is about life. The lives of people and the entire web of life we share the planet with. We need to remind each other how to feel things again. We need to reinstate a culture of care.

Some people will feel inspired to make an artwork, regenerate their local ecosystems, or shine a light on the truth through filmmaking or photography. Others might write a song or a PhD thesis that changes the world. We each have a unique role to play; that's what makes this moment so exciting. The door is open to everyone, no matter their location, age, class, race, gender or education. We can all do what we can to help drive change in our communities, wherever we live in the world. Now, more than ever, it is important to keep our pessimism and optimism in balance: we need to be very clear about the reality of the challenge we face, while being careful not to take on too much negativity, as it smothers the fragile shoots of hope. We can focus instead on regeneration, on the possibility of good, rather than the inevitability of evil. We need our optimism to connect us with a sense of hope, which is fueled by our fellow human beings. At the end of the day, we are relational beings that filter life through our emotions. In the words of American writer and activist Maya Angelou: "People will forget what you said, people will forget what you did, but people will never forget how you made them feel."

*   *   *

Perhaps most importantly, art gives us other ways of experiencing the world that bypass the channels of logic and reason, taking us straight into the domain of the heart. Art has a way of getting us in touch with our common humanity, connecting us with a universal language that transcends our differences. It reawakens our senses. As a climate scientist, it is incredibly exciting to realize that creative communities around the world are now showing up. Some are loud and visible – like the provocative street art and protests we saw at COP26 – and others are quieter, slowly weaving their way into our communities in more local, less visible ways. The beauty of this is that there is a role for everyone. We all have our own unique way of expressing ourselves, and are members of different communities around the world. Art can transform our feelings of grief and anger into beauty and power. It allows us to reframe challenges and consider things from a different perspective, connecting us with others on a deeper, emotional level.

If we place culture at the forefront of our approach to addressing climate change, then the social change we need will start to emerge. Our creative industries have the power to influence how we feel, what we wear, what we eat, what we put in our homes and, ultimately, how we vote. How we act as individuals contributes to the collective culture we experience in our communities, which in turn reinforces what we jointly find acceptable and what we don't. When more and more people get behind new visions of how our society could be, the status quo begins to crumble. The social license for things like polluting our air, destroying our forests and eating animals will start to change when enough people align their actions with their ethics. And when they do, visionary businesses will pivot to meet that demand. We all have power in the consumer and democratic decisions we make; it all adds up to forming a critical mass of progressive people who will eventually nudge others along as our cultures and societies continue to evolve.

Recent research has shown that most people end up siding with the status quo, whatever it may be. Just like incremental changes in natural systems that eventually lead to a rapid transition into a new state, social tipping points can also arise when a critical threshold is reached. When a certain proportion of society changes their views, it eventually has a domino effect on other people. As George Monbiot puts it: "Other people sense that the wind has changed, and tack around to catch it." As we saw in the MeToo movement highlighting sexual violence against women, and the Black Lives Matter protests against racially motivated violence, public opinion can change very quickly when people are outraged enough to remove the tolerance of sexism or racism in our communities. That's how we change our culture: by removing the social license for destructive behaviors that harm other people and the planet, one person at a time.

A 2018 study led by Damon Centola from the University of Pennsylvania in the United States showed that the threshold for a social tipping point was passed when the size of a committed minority reached approximately 25 percent of the population. When this happened, the minority group succeeded in altering the established social convention; in other words, a large enough critical mass was reached that eventually tipped the system. The study highlighted that once this social tippingpoint was reached, between 72 and 100 percent of people eventually adopted the new convention, replacing previously established social norms. They also note that it only takes a small group of ordinary people with a regular amount of social power and resources to successfully initiate social change. Change didn't come about from their authority or wealth, but rather from their passion and commitment to the cause.

Another fascinating study, led by Ricarda Winkelmann from the University of Potsdam in Germany, went one step further and showed that the Fridays for Future climate protests could trigger this kind of social tipping point. The study describes how Greta Thunberg's 2019 school strike snowballed into a social movement that led to unprecedented electoral results for Green parties in the 2019 European Parliamentary Elections, as well as in federal elections in Austria, Belgium and Switzerland. The authors write:

> These bottom-up movements could push the European political system towards a critical 'state', creating the conditions for a tipping process towards radical policy change, ultimately bringing European climate policy in line with the Paris Agreement.

They argue that the European political system is close to a social tipping point, one that will accelerate the political response needed to tackle climate change. Although the momentum of the movement has been slowed by the coronavirus pandemic, the power of the youth activists gathered in Glasgow for COP26 gives me hope that the critical mass we need may not be very far off. It's really inspiring to see new research emerging that shows that a small group of committed people with modest resources really can make a difference. These studies provide empirical evidence to support a long-held belief of those people who have spent a lifetime engaging in social movements. This concept is most clearly articulated by the anthropologist Margaret Mead: "Never doubt that a small group of thoughtful, committed citizens can change the world; indeed, it's the only thing that ever has." In her electrifying book *Hope in the Dark*, Rebecca Solnit writes:

> Sometimes a few passionate people change the world; sometimes they start a mass movement and millions do; sometimes those millions are stirred by the same outrage or the same ideal, and change comes upon us like a change of weather. All that these transformations have in common is that they begin in the imagination, in hope . . . hope is not like a lottery ticket you can sit on the sofa and clutch, feeling lucky . . . hope is an axe you break down doors with in an emergency; because hope should shove you out the door, because it will take everything you have to steer the future away from endless war, from the annihilation of the earth's treasures and the grinding down of the poor and marginal. Hope just means another world might be possible, not promised, not guaranteed . . . to hope is to give yourself to the future, and that commitment to the future makes the present inhabitable . . . wars will break out, the planet will heat up, species will die out, but how many, how hot and what survives depends on whether we act.

There are powerful, altruistic forces already at work in the world, striving to be a part of the rebirth of faith in a better future. Despair is a loss of belief that the struggle is worthwhile. If you are reading this book, you know that giving up on our planet and future generations is simply not an option. You only need to look around you to know that there is still so much worth saving. All of the social justice movements throughout history remind us that we can exercise our democratic rights as citizens to make our voices heard. We must continue to inspire others to swell our ranks until our collective goodwill eventually solidifies into a sense of global citizenship, turning all our local struggles into a united force for justice and hope in this world. It will be a victory wrenched from the vice of despair.

It is incredible to realize that we are living through an era when profound transformation is happening before our eyes; we are living through history in the making, a time that will be talked about for generations to come. It is a truly global movement that will transform life on our planet; what a miraculous time to be alive. And the best part is that there is room for everyone – there is no one way of doing things; whatever you contribute will make a difference. So instead of locking ourselves up in isolation with our own private despair, we can choose to live out our hope and resistance together in public with people of all kinds and forge a sense of global community that seeks to bridge our divides. History is full of examples that show us that it is possible to unite behind our shared values to face the climate crisis together.

We can be part of the generation that chooses to leave a legacy of care and repair; one that sees the Earth not only survive, but thrive. We are each a part of an eternal evolutionary force that has always strived for justice, peace and equality in our communities. We can each do what we can to influence the cultural change that we need to bring about true political leadership. If we are willing to be part of the 25 percent needed to tip the system, then there is a very real possibility that we will transform our world. As South African freedom fighter Nelson Mandela once said: "It always seems impossible until it is done."

12

Revolutions seem impossible
until they become inevitable

I'M SURE SOME OF YOU are wondering – even with all the political will in the world – if decarbonizing the global economy is technically possible? Can we restore our natural ecosystems to their former glory, or is the damage irreversible? The good news is that many of the solutions we need to address climate change are no longer utopian fantasies but are ready to be rolled out right now. If we really put our minds to this, the goal of sustainability could be achieved within a generation.

That's not to say it will be easy, but it is definitely possible. The Montreal Protocol – the international treaty that banned ozone-depleting chemicals – not only reversed the hole in the ozone layer, but also clearly demonstrated how all 198 United Nations member states worked together to deliver what is regarded as the world's most successful environmental agreement. It showed that humans are entirely capable of uniting to protect our planet. The main obstacle to addressing climate change is the lack of political will, not a shortage of readily deployable solutions. As someone who works at a university, I can assure you that the world's brightest minds in the fields of science, engineering and technology are stepping up to solve some of the most difficult technical challenges the world faces. It never ceases to amaze me just how endlessly ingenious humans really are. It's not a story we hear much about, which is why I now want to provide you with some highlights of the inspiring advances that have been made in recent years around the world to accelerate the clean energy revolution and restore life on our planet.

Around 70 percent of carbon dioxide emitted since the start of the Industrial Revolution is from the burning of fossil fuels, with the remaining 30 percent coming from land use changes like deforestation and the degradation of ecosystems. Step one is accepting that the age of fossil fuels has come to an end. The sooner we acknowledge that we need to switch the tap off and start mopping up, the better, as we have a lot of work to do. Currently we burn coal, oil and gas for electricity generation, transport and industrial processes like the production of steel and aluminium. We need to immediately phase out traditional fossil fuels and rapidly scale up renewable energy sources like solar, wind, hydropower and bioenergy, which currently only make up less than a third of global electricity generation.

According to the International Energy Agency's 2021 *World Energy Outlook*, coal is responsible for around three quarters of the global electricity sector's carbon dioxide emissions that have raised global average temperatures over 1°C since pre-industrial times. It's clear that the main source of the problem is how we power our world. This means that how we generate energy has to be at the heart of the solution to bring emissions down. Not only do we need to keep fossil fuels in the ground, but we must phase out existing production and switch to renewable power immediately. Once we do this, we can use clean power to electrify other energy-intensive areas of transport and industrial processes, which could be replaced with low-emissions fuels like hydrogen in the future.

It's really inspiring to realize that the technology needed to achieve the necessary deep cuts in global emissions by 2030 already exists, and the policies that can drive their deployment are already proven. To reach the Paris Agreement targets, the world needs to reach net-zero emissions by 2050. To put us on track to achieve this, we have to see a massive expansion of renewable energy during the 2020s. Although it sounds like a big job, we often don't hear that the world already has more than enough renewable power to meet the energy demands of every person in the world without any shortfall in global energy generation. Conservative estimates that take into account environmental safeguards, land constraints and technical feasibility suggest that solar and wind can supply the world's energy demands more than fifty times over. Other studies claim the number might be closer to 100 times global demand. These are extraordinary statistics to consider – the power of renewable energy is just waiting to be unleashed. Right now, we use 0.01 percent of the available solar potential and 0.16 percent of the wind potential that is available on Earth. According to Carbon Tracker's *The Sky's the Limit* report, just 0.3 percent of the world's land surface needs to be covered in solar panels to provide humanity with all of its energy. That's less than the space taken up by fossil fuel infrastructure today.

Until recently, solar and wind power were more expensive than fossil fuel–derived energy. Almost all solar required economic subsidies to make it competitive with coal, oil and gas, resulting in higher prices. But over the past decade, the cost of solar panels and wind turbines has declined, making the prospect of rolling out renewable energy infrastructure increasingly cost-effective. In 2020, solar power was cheaper than electricity generated by fossil fuels over 60 percent of the Earth's land surface. This means that in vast tracts of the world, clean renewable energy sources are the cheapest form of electricity – not by 2030 or 2050, but *right now*. This figure is predicted to rise to over 90 percent by the end of the decade, as the cost of clean energy continues to come down. According to Carbon Tracker, solar costs have fallen by an average of 18 percent every year over the past decade. At the current 15–20 percent growth rates of solar and wind, fossil fuels will be pushed out of the electricity sector by the mid-2030s, and out of total energy supply altogether by 2050. As the fossil fuel industry will not be able to compete with the phenomenal growth of renewables, the demand for polluting power will inevitably decline, heralding the end of the dirty fossil fuel era and the dawn of the clean energy revolution.

To achieve net-zero emissions by 2050, we must see an unprecedented investment in clean energy. The share of renewables in global electricity generation needs to increase from the current 29 percent to over 60 percent by 2030 and to nearly 90 percent by 2050. To achieve this, we will need to see a five-fold increase in current wind and solar power between 2030 and 2050. According to the International Energy Agency, funding needs to increase to around US$4 trillion annually by 2030, more than tripling current levels. Although mobilizing such a large investment will be challenging, it will provide business opportunities along the entire clean energy supply chain. Everything from equipment manufacturing to deployment and battery storage will generate even more momentum in the rollout of clean energy, driving the costs down even further. While the level of investment might seem large, remember that in 2020, governments around the world spent US$450 billion in direct subsidies to the fossil fuel industry to keep polluting the planet. Imagine what we could achieve if we redirected these funds into the renewable energy sector, improving efficiencies in some of the remaining bottlenecks of battery storage, integration of renewables into the electricity grid and the construction of long-distance transmission lines. With the right government policies and regulatory frameworks, the business sector will move its money away from the fossil fuel industry and instead invest heavily in the technology and infrastructure needed for the clean energy transition, creating new industries that will employ millions of people around the world.

As the coronavirus pandemic showed us, countries can move fast to deal with a crisis when they have to. For example, as of December 2021, the United States government had approved US$4.5 trillion of spending to deal with the pandemic since early 2020, with US$3.5 trillion spent so far. By mid-2021, over US$16 trillion had been spent globally to provide economic relief and deal with the crisis. This level of response clearly shows the ability of governments to mobilize emergency funding. It highlights that if our political leaders really wanted to, they could redirect funds to support renewable energy technology. Which brings us back to social tipping points. Once our communities remove the social license for the continuation of the fossil fuel industry and demand that governments fund the clean energy transition, the future will be back in our hands. All the technology we need exists right now – the only thing missing is the political will, and that's something every single one of us has the power to influence at the ballot box.

*   *   *

The thing I love most about the clean energy revolution is that developing nations are set to be the biggest winners. When you look at a global map of solar power potential, the so-called "sun belt" stretches from the equator to 35 degrees north and south. It takes in tropical and subtropical regions that contain around 70 percent of the world's population, in places like Africa, Latin America, Asia and the Middle East. Investing in renewable energy infrastructure in these places will allow countries like China, India, Mexico and Morocco to become clean energy "superpowers" that will be able to export excess energy to parts of the Northern Hemisphere that have more limited solar potential. The more I think about it, the more it feels like poetic justice – the developing world will finally have the advantage, which could see a realignment of geopolitical power away from the colonial powers of the past in our redefined future.

Africa is one of the continents that has a lot to gain from the renewable energy revolution. It is estimated that if all the sunlight received in North Africa could be converted into solar energy, it would be enough to power all of Europe more than 1000 times over. With the Sahara Desert averaging around nine hours of sunlight each day, Africa could be transformed into a solar energy powerhouse, with nations like Tunisia, Egypt and Morocco generating electricity that could support millions of homes in Europe and the Middle East. And as a bonus, researchers have found that the installation of large-scale solar and wind farms in the Sahara could double the amount of local rainfall, especially in the Sahel. Their model analysis found that a combination of increased surface drag and reduced reflectivity promotes the growth of vegetation, which in turn further increases rainfall. So not only would renewable energy reduce carbon emissions and generate cheap electricity for the region, but it might also help ease water scarcity issues. It really doesn't get much better than that.

A comprehensive analysis of solar photovoltaics by Marta Victoria from the Aarhus University in Denmark clearly shows that it is a mature technology that could be quickly rolled out to produce most of the world's energy in the crucial period between 2030 and 2050, when we need to reach net-zero emissions. The study explains that global solar electricity generation has grown 50 percent each year since 2008 and that this trend is expected to continue, especially in China and Europe. Exciting advances in solar cells like bifacial panels, which use both sides to collect sunlight, and the development of tandem solar cell systems that can be easily modularized and rapidly deployed from a single household to an industrial scale, mean that the technology needed to power our sustainable future is well and truly here. It just needs political support to be scaled up and rolled out all over the world.

Another advantage is that many renewable energy jobs will be generated in rural and regional areas that have the physical space to support large-scale projects, which will help reduce entrenched inequalities that exist between city and country populations. This is particularly important in regions that are currently heavily dependent on fossil fuel industries like coal mining and gas exploration. Ensuring that these workers have opportunities to retrain and transition into new renewable energy jobs will ensure that regional communities won't be left behind when the coal mine that might have employed their families for generations finally shuts down. While the fossil fuel industry powered the nineteenth and twentieth centuries, renewable energy is clearly the way of the future. There's no reason why these communities can't be an integral part of transforming the world.

The good news is that the number of renewable energy jobs around the world is growing rapidly. According to the International Renewable Energy Agency, there were 12 million people employed by the renewable energy sector in 2020, with 39 percent of those jobs in China alone. Other places like Brazil, India, the United States and the European Union also employ many people in the sector. The solar photovoltaics industry employs 4 million workers, representing a third of total renewable energy jobs. As more and more investment is directed into renewable energy projects, new job opportunities will be created in the technologies needed to power the future.

Like all things in life, technology eventually moves on. You only need to think back to old VHS tapes, photographic film or mobile phones from the 1980s to know that you can't stop technological progress. Human ingenuity and innovation will always strive to improve the tools we have. So while we often hear negative stories about the end of the fossil fuel industry, perhaps we just need to reframe things. Instead of buying into the scaremongering that the global economy will collapse if we transition away from polluting industries, we can think of this moment in our history where we finally figured out how to live sustainably on our planet.

But make no mistake: disentangling the fossil fuel industry from politics all over the world is going to be one of the fiercest fights of all time. In "Big Daddy Gas," a chilling critique of the destructive industry's grip in Australia, one of Australia's finest writers, Tim Winton, likens the struggle to trying to leave an abusive relationship:

> Our democracy has been so bewitched, and so thoroughly gaslit, that the nation is now terrified at the prospect of leaving what's clearly a toxic relationship. We know it's bad for us and for the kids. But we've become so worn down and disoriented, so lied to and loomed over that we're scared to get out. And those to whom we look, for guidance and support, the folks who should be finding us a pathway out of this mess? They're more rattled than us. Because they know Big Daddy. What he's capable of. They insist he's a top bloke – mostly. Just gets a little carried away, sometimes, that's all. When we cry out to be rescued, they're keen to remind us about how generous Big Daddy's been. He's been so nice to us, hasn't he? He loves us. In his way.

No matter what distorted reality gets thrown at us as we head for the door, we must get out. Another world is possible.

Decarbonizing the global economy will be a monumental leap forward for humanity, one that will be talked about by historians in the future. We are the generation living through this time of transition, so it will be hard for some people currently employed in the fossil fuel sector. But with enough visionary planning, our communities and economy will not only survive, but thrive through this era of rapid change.

*   *   *

You don't have to look too far to find inspiring stories of the individual countries, states and local communities already embracing the clean energy revolution. There are now sixteen countries that generate

90 percent or more of their electricity from low carbon sources, with all having some reliance on hydropower or nuclear energy. They include Iceland, Ethiopia, Bhutan and Norway, which rely heavily on hydropower, and Switzerland, Sweden and France, which use a lot of nuclear power. While they have low carbon footprints, critics of these technologies cite other considerable issues associated with them, namely the enormous cost of constructing and maintaining nuclear and dam facilities, and their significant environmental impacts. These include the accumulation of radioactive waste, the catastrophic societal risks associated with structural failures or industrial accidents, the potential proliferation of nuclear weapons, the drowning of natural landscapes and Indigenous lands, and the displacement of human and ecological communities. As they are very expensive infrastructure projects, they are often propped up by government subsidies that divert public investment away from the development of solar and wind projects that would provide highly decentralized forms of electricity that literally place power back in the hands of local communities.

Denmark is a country attempting to decarbonize without either of these sources. It has vast fossil fuel resources yet is showing the world what can be achieved when a federal government encourages the growth of the renewable energy sector with smart industrial policies. Driven by public investment, the country has become a key developer of wind farms and is home to major turbine manufacturing and leading research and development centers. Danish companies are now estimated to own 40 percent of the European offshore wind market, creating jobs mostly in remote coastal communities, cementing its position as a world leader in renewable energy technology. It's quite a departure from history: the port of Esbjerg was once Denmark's leading service hub for the oil and gas industry, before transforming itself over the past two decades into the world's largest center for offshore wind. These days, 25 percent of Esbjerg's revenue is generated from offshore wind power, compared with just 10 percent from oil and gas. The port is currently involved in more than fifty wind projects, representing around half of Europe's total wind capacity, making it one of the biggest renewable energy hubs in the world.

In 2021, around 65 percent of Denmark's power generation came from renewable energy sources, mostly from wind. The nation has legislated impressive targets for reducing greenhouse gas emissions, aiming to be 70 percent below 1990 levels by 2030 and to reach carbon neutrality by 2050. As part of their policy to phase out fossil fuel extraction, they also plan to end all new oil and gas exploration in the North Sea by 2050. These are very impressive goals for the European Union's largest oil and gas producer (which excludes the non-EU nations of Norway and the United Kingdom) to take. Such strong action will transform Denmark from a country historically reliant on fossil fuels to a nation powered by clean energy. With luck, their visionary action will inspire other countries to take their own bold steps towards making the decarbonization of the global economy a reality.

Unlike progressive Denmark, Australia's federal government plans to continue expanding the fossil fuel industry, even though the nation has solar and wind power potential that is the envy of the world. According to Carbon Tracker's *The Sky's the Limit* report, Australia has the highest renewable energy potential per person in the world, placing the nation in a "league of its own" to become "the battery of the world." Australia's solar and wind potential is estimated to be over 10,000 megawatts per person per year, which is at least five times greater than the other high potential regions of South America, Africa and the Middle East.

In his visionary books *Superpower* and *Reset*, esteemed economist Ross Garnaut makes a solid case for Australia's potential to be a clean energy superpower. He argues that the country has unparalleled natural resources and technological skills that could see the nation reach

100 percent renewables in little more than a decade and position itself to be a global leader in the clean energy economy. And yet, currently 62 percent of Australia's electricity generation comes from coal and a further 10 percent from gas, with alarming plans to develop 116 new fossil fuel projects. This is despite the fact that the government's own research by the CSIRO, Australia's national science agency, has reported that renewables are the cheapest form of new electricity generation, even factoring in costs associated with storage and transmission.

Despite the shameful lack of leadership at the federal level, Australia's renewable energy revolution is being led by regional state and territory governments. In 2020, 28 percent of the nation's electricity came from clean energy sources, with wind and rooftop solar leading the way. The renewable energy industry currently employs around 25,000 people, with existing projects in the pipeline set to create another 20,000 jobs if they can secure government backing. In 2020, the island of Tasmania became the first Australian state to generate 100 percent of its electricity from renewable sources, mainly hydropower. The state of South Australia passed a significant milestone in October 2020, when 100 percent of its electricity came from solar for one hour, the first time this had been achieved by a jurisdiction of its size anywhere in the world. Over the entire course of 2020, 60 percent of South Australia's electricity came from renewable energy. Then in the final days of 2021, South Australia set another impressive new record, with the state's solar and wind farms and rooftop solar systems supplying over 100 percent of local demand every day for a period of one week. Clearly the region is on track to be a renewable energy powerhouse. There are plans for South Australia to generate 500 percent of the state's electricity demand by 2050, setting itself up to become a national and international exporter of clean energy. This is likely to be one of the most ambitious renewable energy targets anywhere in the world.

The rooftop solar industry has experienced phenomenal growth in recent years; now more than 3 million Australian households and small businesses have solar power installed. This makes Australians world leaders in the uptake of rooftop solar, representing around one in four households around the country. Uptake has been particularly impressive in South Australia, where experts believe that rooftop solar will be able to supply 100 percent of the state's energy needs in the near future. It's a wonderfully subversive trend that suggests that the federal government's pro-fossil fuel stance is out of touch with the community's strong support for clean energy. Not only are they keen to reduce their power bills, but many people are motivated by a desire to do what they can to address climate change.

In the regional town of Yackandandah in northeast Victoria, locals banded together to purchase a community-owned battery that allows many households to share their excess solar power with their neighbors. Instead of each home installing their own private battery, which are still prohibitively expensive for most people, they chose to work together to create a microgrid for the town that allows everyone to access stored energy at night or when the sun isn't shining. People without solar panels can choose to purchase power through their community-owned energy retailer, Indigo Power. The goal is to power their town with 100 percent renewable energy by 2022 by pooling their resources and working together as a collective.

The Totally Renewable Yackandandah initiative is an incredibly inspiring example of how a volunteer-run community group, backed by visionary state government initiatives and local people, can work together to help switch to renewable energy one town at a time. We just need our federal government to step in and support the acceleration of the renewable energy transition being led at the regional level. There needs to be a co-ordinated plan to ensure we develop national infrastructure and systems that can deal with the transition. What's happening at the grassroots across Australia gives me hope that the clean energy revolution is now not only possible but inevitable. Despite the federal government's lack of leadership, millions of people across the country are stepping up to light the path to our clean energy future. And if this can happen in Australia – a country still under the spell of the fossil fuel industry – it can happen anywhere.

*   *   *

While the renewable energy transition is certain, the restoration of the world's ecosystems is going to be more difficult, as it is intertwined with global food production and the value we place on our natural environment. It's going to require a major rethink of humanity's role on the planet. As well as replacing the burning of fossil fuels with renewables, we also need to turn our attention to the remaining 30 percent of carbon emissions that come about from altering the land surface through processes like deforestation. As Paul Hawken outlines in his inspirational book *Regeneration: Ending the Climate Crisis in One Generation*, there have been many efforts by local communities around the world to implement nature-based solutions that look to restore natural carbon sinks by protecting forest and wetland ecosystems and by implementing regenerative agricultural practices that lock up carbon and methane in natural environments. Hawken writes: "When life regenerates, complexity proliferates. Diversity burgeons. Productivity soars. Species reappear. And the climate responds."

But despite impressive community efforts to restore ecosystems, the IPCC's *Mitigation of Climate Change* report says that past policies applied to the agriculture, forestry and other land use sector has so far only reduced total global emissions by 1.4 percent over the 2010–2019 period. Clearly there is a very, very long way to go. Given that changes in land use are required to deliver around a quarter of the pledged emissions reductions needed by 2030 as part of Nationally Determined Contributions under the Paris Agreement, it is crucial that genuine progress is made over the coming decade if we are to try and limit warming to below 2°C. The good news is that mitigation efforts in this sector will also help address the related issues of biodiversity loss and food sustainability, which will improve both human and planetary health.

One of the most positive outcomes of the COP26 meeting was the signing of the Glasgow Declaration on Forests and Land Use, which was an attempt to revive the 2014 New York Declaration on Forests that sought to end deforestation and eliminate the practice from agriculture by 2030. One hundred and forty-one nations containing 90 percent of the world's forests agreed to work collectively to end and reverse forest loss and land degradation by 2030. Some experts hailed it as the "Paris moment" for forests. The move was a critical step towards uniting behind the goal of achieving sustainable land use practices, allowing the Earth's ecosystems to naturally absorb as much carbon from the atmosphere as possible. The declaration acknowledges the interconnectedness of climate and biodiversity and the importance of achieving sustainable land use practices to maintain critical ecosystem services and functional habitats to support the diversity of all life on our planet. Significantly, the signatories included Brazil, Indonesia and the Republic of the Congo, which house the largest remaining tracts of rainforest in the world. It was an important diplomatic step that hopes to reinstate humanity's role as custodians rather than conquerors of nature.

But as always, the devil is in the detail; concrete plans and accountability measures need to be in place to ensure progress is made beyond just good intentions. In places like Brazil, Indonesia and the Congo, forests are cut down to make space for animals to graze or for growing large agricultural crops. The financing for activities that destroy natural ecosystems is nearly forty times greater than money spent to protect them. According to *Forbes* magazine, US$26 billion has been invested since 2010 to save forests, compared with the US$1.3 trillion that has been spent on clearing them. The Glasgow forests pledge included over US$19 billion in public and private funds to address this, with fourteen countries and philanthropic donors also pledging at least US$1.7 billion from 2021 until 2025 to advance Indigenous peoples' and local communities' land rights, and support their role as guardians of forests and nature. Importantly, CEOs from more than thirty of the world's largest financial companies, including Axa, Aviva and Schroders, have committed to end investment in activities linked to deforestation. Governments of twenty-eight countries also committed to the removal of deforestation-related agricultural commodities including palm oil, soy and cocoa, which is a significant task that will involve forest monitoring and supply chain traceability. There is no denying that policing the forestry industry will be difficult and politically sensitive, but it is an essential step in trying to restore balance to our planet's ecosystems and climate.

One of the most important aspects of trying to reduce deforestation in places like the Amazon is making the connection between agriculture, environmental destruction and climate change. According to the IPCC's *Sixth Assessment Report*, in 2019 agriculture, forestry and other land use contributed to almost a quarter (22 percent) of total global anthropogenic emissions of carbon dioxide, methane and nitrous oxide mainly related to deforestation and agricultural practices used in food, animal feed and timber production. This is a concern, as natural land processes currently absorb almost a third of carbon dioxide emissions from fossil fuel burning. Unlike greenhouse gas emissions generated by other sectors like energy, industry, transport and the operation of buildings, land use emissions are typically higher in developing countries compared to developed nations. In Africa, Latin America and Southeast Asia, carbon dioxide emissions from land clearing dwarf those from other sectors, making up more than

50 percent of emissions in these regions, related to the expansion of agriculture into carbon-dense tropical forest regions where vast quantities of carbon dioxide are released due to the removal and burning of biomass and the draining of carbon-rich soils. Since 2000, land use emissions have risen faster than fossil fuel contributions, with more methane released through the alteration of peatland and wetlands ecosystems and the expansion of livestock and rice cultivation.

The IPCC estimates that global food production currently accounts for around a third of total greenhouse gas emissions. The food system takes up a staggering half of the Earth's habitable land, with the vast majority of it used for livestock and their feed. Over the past sixty years, the number of calories we consume globally has increased by a third, with meat consumption more than doubling to 43 kilograms per person each year. Just as richer countries tend to have higher greenhouse gas emissions, they also tend to eat more meat, with the average American and Australian consuming around 120 kilograms of meat each year compared with 4 kilograms consumed by the average Indian. To make matters worse, 25–30 percent of the total food produced around the world is lost or wasted. These confronting trends have resulted in immense pressure on natural ecosystems to make way for crops and grazing land, leading to a disastrous loss of biodiversity and the degradation of ecosystems.

In particular, the meat industry has the most destructive impact on our natural environment. According to a recent study led by Xiaoming Xu from the University of Illinois, the greenhouse gas emissions needed to produce animal-based foods are twice those associated with plant-based foods. Cows, pigs and other animals that humans eat account for a shocking 57 percent of all emissions associated with food production. Beef and dairy (cow milk) have the highest footprint, together accounting for 45 percent of total animal-based food emissions. Beef alone accounts for a quarter of emissions produced by raising and growing food and is the number-one driver of deforest-ation of the world's tropical forests in places like Brazil, Argentina and Paraguay. This is because livestock requires a lot of land for grazing, which is often acquired by bulldozing forests. Vast tracts of land are also needed to grow their feed crops of maize, wheat and soybeans. As if that wasn't enough, livestock also produce large quantities of methane – a greenhouse gas thirty times more powerful than carbon dioxide – further amplifying warming.

The difference in emissions between meat and plant production is stark. Xu estimates that 2.5 kilograms of greenhouse gases are emitted for every 1 kilogram of wheat produced. In contrast, a single kilo of beef creates approximately 70 kilograms of emissions – that's nearly thirty times the greenhouse gas footprint of wheat. Their research clearly shows that raising and slaughtering animals for food is far worse for the climate than growing and processing fruits and vegetables for people to eat. In a study full of astonishing results, perhaps most surprisingly the authors find that the majority of the world's croplands are used to feed livestock rather than people. That's staggering logic to try to take in – instead of using food to feed people, we are feeding it to animals, which we then eat. Given that around a third of total greenhouse gas emissions comes from food production, we need to transform indus-trial agricultural practices and human diets if we are to limit global warming. If we are serious about addressing climate change, then it is time for a major rethink of our eating habits and the farming practices needed to sustain our food choices.

Moving away from eating lots of meat to adopting a plant-based diet is something many healthcare professionals support. In 2019, Walter Willett from Harvard University led a comprehensive study published in the medical journal *The Lancet* that brought together experts from sixteen countries in various fields of human health, agri-culture, political sciences and environmental sustainability to develop science-based guidelines to identify the characteristics of a healthy diet and sustainable food production. This formed the basis for the development of the EAT-Lancet Commission's recommendation of a plant-based "planetary health diet," which largely consists of vegetables, fruits, whole grains, legumes, nuts and unsaturated oils, low to moderate amounts of seafood and poultry, and no or very little red meat, processed meat, added sugar, refined grains and starchy vegetables. The diet recommends no more than three small portions of meat, two serves of fish and seven glasses of cow milk each week to minimise your environmental footprint.

Willett's team explains that food production is among the largest drivers of global environmental change by contributing to climate change, biodiversity loss, freshwater use, interference with global nitrogen and phosphorus cycles and land use change. They make the case that adopting a plant-based diet is not only far healthier but also more environmentally sustainable, concluding that the world needs "nothing less than a Great Food Transformation" to deliver a win-win for people and the planet. It's a conclusion also reached independently by the IPCC in their *Special Report on Climate Change and Land*:

> Balanced diets, featuring plant-based foods, such as those based on coarse grains, legumes, fruits and vegetables, nuts and seeds, and animal-sourced food produced in resilient, sustainable and low-GHG [greenhouse gas] emission systems, present major opportunities for adaptation and mitigation while generating significant co-benefits in terms of human health.

According to a 2018 study led by Joseph Poore from the University of Oxford, on average, emissions from plant-based foods are ten to fifty times smaller than those from animal products. These figures clearly show us that the food we eat is having a very heavy footprint on the planet. We need to move away from dietary choices that require more land and water and cause more emissions of planet-cooking gases. To achieve this transformation of the global food system we need individual consumers, policy makers and everyone in the food supply chain to work together towards the shared global goal of healthy and sustainable diets for all. This will provide an opportunity for farmers to use regenerative agricultural practices to restore degraded land and soils that will not only rebalance our food system but also absorb greenhouse gases needed to stabilize planetary warming.

Once again, the science tells us that the solutions we need to transform our planet are already here. Reducing the impact on our ecosystems by modifying what we eat is something we can choose to do today. We can remove the social license for high-impact, industrial agriculture, one meal at a time. You don't necessarily have to become a vegan or vegetarian (although that would be ideal), but it wouldn't hurt to think about adopting a planetary health diet that significantly reduces the amount of meat and dairy you consume. Given that the natural world is in desperate trouble, perhaps it's time to extend the care and respect we show for the pets in our lives to all other life forms we share the planet with. We can choose to show compassion for the domesticated animals that we kill for food and the wild creatures we displace from the world's forests, and still be well nourished and healthy. The choice of what we eat is literally in our hands. We just need to put our money where our mouth is.

*   *   *

The clearing of land for food production and our exploitation of wildlife also has deadly consequences for human health. Research has shown that around 70 percent of emerging infectious diseases, and almost all recent pandemics, originated in animals. Nearly three quarters of these came from wild animals, with complex interactions between domestic animals and humans. The emergence of infectious diseases is correlated with human population density and wildlife diversity, and is driven by human-caused land use changes such as deforestation, the expansion of agricultural land, intensive livestock production, and the increased harvesting of wildlife for "bush meat." These processes disrupt ecological systems, which in turn influences the risk of pathogens that cause infectious diseases in humans, particularly in the biodiverse tropical regions of Africa and Asia.

Protecting intact forests not only benefits biodiversity conservation and global carbon storage, but also prevents the risk of disease transmission. The encroachment of human activities into natural areas leads to closer interactions between wildlife, livestock and people, increasing the chance that an infectious disease will jump the species barrier into humans. This was the case for Nipah virus in Malaysia in 1998, Severe Acute Respiratory Syndrome (SARS) in China in 2003, and Ebola in West Africa from 2013 to 2016. In these cases, the emergence of Nipah virus was linked to the intensification of pig production at the edge of tropical forests where infected fruit bats live, while the SARS and Ebola viruses were traced back to areas where bats are either hunted or under increasing pressure from human development. There are also horrifying examples of diseases emerging through the use and trade of wildlife, like the association between the consumption of chimpanzee and HIV/AIDS, the link between SARS and markets containing wildlife, and the outbreak of Ebola with the hunting of great apes, including gorillas.

Although the precise origin of COVID-19 remains uncertain, analysis of genome sequences has demonstrated that it is a zoonotic disease – an infection that passes from animals to humans – that likely originated in the Huanan market in Wuhan, China, where the first cases of the disease were reported. The live animal and seafood market is known to contain a lot of exotic wildlife stacked in crates and cages for sale, including species that are known to host a range of infectious diseases.

A study by Kristian Anderson from the Scripps Research Institute in the United States has demonstrated that the genetic material contained in the SARS-CoV-2 virus that causes COVID-19 is 96 percent identical to the SARS coronavirus, which originated in bats. Anderson's research shows that while Malayan pangolins illegally imported into the Guangdong province of China also contain coronaviruses similar to SARS, the bat virus is the closest genetic match to the SARS-CoV-2 sequence.

Emerging evidence based on an analysis of the original Wuhan outbreak suggests that racoon dogs, which contain coronaviruses also linked to the outbreak of SARS, may also be involved in the transmission chain. A 2021 study by Michael Worobey from the University of Arizona reports that live mammals susceptible to coronaviruses, including raccoon dogs, were sold at the Huanan market and three other live animal markets in Wuhan before the pandemic. He explains that the earliest symptomatic cases of COVID-19 were linked to the western section of the market, where racoon dogs were caged. Whether the disease passed directly from bats to people, or through an intermediate host like racoon dogs, remains a mystery. Unfortunately, no live animals collected from the markets were screened and the Huanan market was closed and disinfected on 1 January 2020, so we may never know which infected animal was responsible for unleashing this lethal disease on the world.

Whatever the precise origin of the COVID-19 vector might be, it is clear that our ongoing abuse of nature will continue to devastate humanity. In May 2022, the World Health Organization estimated that the COVID-19 pandemic had killed over 15 million people worldwide, with economic damage estimated to be in the trillions. Lives everywhere have been indefinitely upended as we struggle to live through an ongoing public health emergency. Where this will all end is unknown, but what is clear to scientists is that the emergence of COVID-19 is yet another SOS signal from the natural world. If we fail to rebalance our relationship with nature, the environmental, health and economic impacts will be brutal. If we begin to join the dots between our actions and their consequences, we can avoid further nightmare scenarios like the COVID-19 pandemic, and begin to repair our relationship with all life on Earth.

*   *   *

Humanity's destruction of the natural world is now so great that we must urgently scale up international efforts to restore natural ecosystems if we are to stabilize the Earth's climate. The climate and ecological crises are one and the same. According to the IPCC, effective conservation of 30-50 percent of the Earth's land, freshwater and marine areas would help protect biodiversity, build ecosystem resilience and maintain ecosystem services that are vital to food production and human health and wellbeing. In an attempt to address the global environmental emergency, in 2021 the United Nations released a draft of the *Post-2020 Global Biodiversity Framework* calling for 30 percent of the Earth's land and sea areas to be conserved by 2030. The initiative is being led by the High Ambition Coalition for Nature, an intergovernmental group of seventy countries with the goal of restoring the natural world. Currently less than 15 percent of the world's land, 21 percent of freshwater systems and just 8 percent of the ocean have some degree of protection, often with insufficient stewardship to prevent further damage or increase resilience against climate change. To give nature a fighting chance, we need to at least double current land protection and more than quadruple efforts to protect the ocean.

This "30 by 30" proposal is seen as a vital step towards the UN Convention on Biological Diversity's goal of "living in harmony with nature by 2050." The vision of the framework is to:

Take urgent action across society to conserve and sustainably use biodiversity and ensure the fair and equitable sharing of benefits from the use of genetic resources, to put biodiversity on a path to recovery by 2030 for the benefit of planet and people.

With effective partnerships between Indigenous people and other local communities, the plan has the potential to finally reinstate humanity's custodianship of the planet. Like all international efforts, these things are never going to be quick or easy, but they are an important starting place for ongoing discussion. When dealing with global-scale problems, it's important to remember that perfect is the enemy of good. If we start from a place of good faith and a willingness to compromise and collaborate, human history tells us that anything is possible.

In 2021, a team of over 100 economists and scientists led by Anthony Waldron from the University of Cambridge produced a comprehensive report on the economic implications of protecting nature. They note that the World Economic Forum now ranks biodiversity loss as a top-five risk to the global economy, arguing the financial case for the expansion of conservation areas to cover 30 percent of the Earth's surface. Their analysis showed that expanding protected areas would generate an extra US\$64 billion per year by 2050, with forests and mangrove ecosystems alone saving US\$170–534 billion each year by protecting societies from impacts associated with flooding, climate change, soil loss and coastal storm damage. The ambitious target would increase the area recognized by Indigenous people and other local communities by 63–98 percent, improving opportunities for intergenerational environmental stewardship to be reinstated and strengthened into the future.

With this comes the opportunity of restoring Indigenous land rights, which would go a long way to repairing the damage done to people displaced from their traditional lands by colonialism and corporate interests. In the book *Regeneration*, Debra Anne Haaland, member of the Kawakia tribe of New Mexico, explains that the success of the "30 by 30" plan depends on:

Stopping attacks on Indigenous land and people. It means supporting efforts to restore cultures that are being subject to racism, that have suffered trauma, that wish to revive their language and reclaim their sovereignty where it has been violated. There is an extraordinary teaching about the Earth that is needed, a way of knowing that erases the separation between people and nature, a disconnection that has caused the climate crisis. That knowledge is here.

Today, around 400 million Indigenous people live on a quarter of the world's land, and inhabit around 85 percent of areas designated for biodiversity conservation. These communities have observed the weather, plants, animals and migrations for millennia, immortalizing them in metaphors and stories that have been passed down in unbroken traditions for countless generations. Worldwide, First Nations people are struggling to save their traditional lands from exploitation that began with colonial dispossession centuries ago. As Haaland explains, the brutal conquest of Indigenous communities was based on the fifteenth-century papal degree that became known as the "Doctrine of Discovery." This meant that land could be claimed in the name of a Christian monarch when a flag was planted in the "New World" lands of Australia, New Zealand, North America, Africa and Asia. Although Pope Francis apologized to all Indigenous people for the "grave sins" of the church in 2015, Haaland notes that the Doctrine of Discovery has never been rescinded by the Catholic Church.

As climate change has worsened, there has been a big push by Indigenous groups around the world to fight for climate justice. They are some of our most powerful leaders in the climate movement because they understand the imbalance we face better than anyone else. They know in their bones what's at stake. In the aftermath of Australia's catastrophic Black Summer bushfires of 2019–2020, Indigenous journalist Loreena Allam wrote for *The Guardian*:

Some of these places have never burned, not once in my lifetime, let alone all at once. Like you, I've watched in anguish and horror as fire lays waste to precious Yuin land, taking everything with it – lives, homes, animals, trees – but for First Nations people it is also burning up our memories, our sacred places, all the things which make us who we are. It's a particular grief, to lose forever what connects you to a place in the landscape. Our ancestors felt it, our elders felt it, and now we are feeling it all over again as we watch how the mistreatment and neglect of our land and waters for generations, and the pig-headed foolishness of coal-obsessed climate change denialists turn everything and everyone to ash.

In response to the escalating environmental and cultural destruction of their lands, in recent years we've seen the Indigenous Environmental Network in North America, the Pacific Climate Warriors from fifteen island nations, and Australia's Seed Indigenous Youth Climate Network, made up of Aboriginal and Torres Strait Islander people, rise up to protect their homes from the impacts of climate change and ongoing efforts to expand the fossil fuel industry on traditional lands. Celebrated Australian Indigenous author Tony Birch has written extensively on the suffering experienced by First Nations people by colonial invasion and the issue of climate change. Birch writes of the need to be willing to start from a place of acknowledging the trauma of our colonial past:

> The relationship between colonialism, capitalism and environmental degradation and a consequent link to climate change is unambiguous. In Australia, the usurpation of land not suited for wide-acre agricultural farming had led not only to the appropriation of Indigenous land, but also the destruction of local ecologies and the wasteful use of natural resources such as water and soil. Jon Altman's assessment that "the brutal colonisation and political marginalisation of

Indigenous Australians can be understood as a conflict over land and resource rights" accurately reflects the extent of violence utilised by colonial forces in an effort to dispossess Indigenous people of country. This is not a conflict located in the past. The consequences of colonialism reverberate in contemporary Australian life. Inequitable socio-political and economic structural frameworks dominate relationships between Indigenous and non-Indigenous Australia, entrenching the marginalisation and disempowerment of Indigenous communities.

Until we are prepared to face these hard truths and shameful collective histories, it will be difficult to make peace with each other and rebalance our relationship with the Earth. In Australia, Indigenous communities often refer to the idea of environmental stewardship as "caring for Country," knowledge that has been passed down from generation to generation for at least 40,000 years. Placing the people who know most about the deep connection between culture and the natural world at the heart of conservation is a profound opportunity for healing ourselves, the land and our oceans. Unless we are willing to recognize how far we have strayed from our inherent humanity, it will be impossible to restore our relationship with the Earth. As Potawatomi Nation writer and scholar Robin Wall Kimmerer from the State University of New York writes in *Braiding Sweetgrass*:

> The consumption-driven mindset masquerades as "quality of life" but eats us from within. It is as if we've been invited to a feast, but the table is laid with food that nourishes only emptiness, the black hope of the stomach that never fills.

Humanity has a lot to learn from First Nations people about how to heal our relationship with the planet. We can refuse to participate in an economy that continues to destroy the natural world for corporate profits, and instead choose to protect what is left and begin reversing the damage. While our landscapes are unlikely to be restored to their former precolonial glory, perhaps a coalition of the willing can be formed between Indigenous people, scientists and local communities to restore our relationships with each other and the natural world we all love so much. It could be the birth of a new generation of conservationists who will do whatever they can to begin healing our world; perhaps there is nothing more important to aim for at this moment in time than this.

Regenerating the natural world will not only absorb enormous amounts of carbon by locking it away in our protected places, but it will also restore degraded ecosystems that will see biodiversity flourish in years to come. You only have to look to the success of the recovery of whale populations in our oceans following decades of anti-whaling campaigning to know that when enough people care, we can act as an international community to restore the natural world. Our small steps now towards repairing at least 30 percent of the Earth's land and ocean ecosystems will eventually be looked back on as foundational efforts for major victories that will be experienced in the future. It could be humanity's most enduring legacy, our ultimate gift to future generations – the restoration of hope and life on our planet. We can choose to make this the moment when we reclaim our rightful place as custodians of nature, living peacefully alongside the stunning array of creatures we share our earthly home with. When we choose to save the planet, it will be a conscious choice to save ourselves.

In *Hope in the Dark*, Rebecca Solnit writes of the immense joy local people experienced as they witnessed the return of hundreds of thousands of sockeye salmon to Lake Washington in Seattle:

> Their return was not . . . the revitalization of an ancient salmon run;
> they were hatchery fish returning to where scientists at the University

of Washington had hatched them. They were no pure ancient past coming back but they were one version of a future with room in it for some kind of wildness ... there is a hope in that for which we might gladly surrender purity.

As the old adage goes, the journey of a thousand miles begins with a single step. Learning to live sustainably on our planet could be humanity's greatest achievement, a redemptive moment for our species that future generations will thank us for. In the words of David Attenborough:

> *Homo sapiens*, the wise human being, must now learn from its mistakes and live up to its name. We who are alive today have the formidable task of making sure that our species does so. We must not give up hope. We have all the tools we need, the thoughts and ideas of billions of remarkable minds and the immeasurable energies of nature to help us in our world. And we have one more thing – an ability, perhaps unique among the living creatures on the planet – to imagine a future and work towards achieving it. We can yet make amends, manage our impact, change the direction of our development and once again become a species in harmony with nature. All we require is the will.

We just need to open our hearts and begin the long journey home together.

Homecoming

IN THE FINAL DAYS OF 2021, I visit one of my favorite places not far from home: the Nightcap National Park in northern New South Wales, where all the things worth fighting for are on spectacular display. Two years earlier, the usually wet subtropical rainforests of Terania Creek were impossibly ablaze. Although the nearby ridgetops were burnt, the lush gorge containing Protesters Falls – named after the iconic forest blockades that gave birth to Australia's environmental movement – were mercifully spared. Although these superb rainforests were saved from loggers by the local community in the late 1970s, some forty years later, their survival is now threatened by worsening heat, drought and fire. But, for now, a second wet summer in a row has brought relief. I take a slow, deep breath and feel my molecules rearranging. The air is different here; its crisp density tingles in my lungs. The sound of the rushing creek beckons me to stop, to listen. So I do, pausing long enough for this primal communion to restore me.

As I walk the winding track towards the waterfall, my weary body absorbs the sensory delights. Groves of Bangalow palms tower overhead, their prehistoric silhouettes ragged against the saturated sky. Enormous staghorn ferns lavishly drape from the canopy, their presence a celebration of the richness of life high above the forest floor. Around me, trees and vines spiral and twist together, inseparable, like a lover's embrace. I slowly follow the creek until the path gives way to a scramble over smooth boulders to reach the waterfall. Its beauty catches my breath: wispy cascades leaping from a sheer, 30-meter cliff above; the pool below surrounded by a natural rock amphitheater. Its colossal sandstone blocks are studded with a stunning array of plants that defy gravity and logic. Life somehow thrives on the edge of things, miraculously and magnificently holding on.

I find a giant boulder to rest on, and lie down to take in the immensity of the landscape. The cool spray of the falls gently mists my face. I close my eyes to soak in the charge of it all. My mind drifts before it settles into a place where stillness is always and already here, awaiting my return. I stay this way until I dissolve into the rock beneath me and merge with the sky above. I quietly pay my respects to all those who have come before me: the Bundjalung custodians of this land, the community who fought to protect this place from intergenerational vandalism, and all those who sit silently here to remember why. When I finally peel my spine from the boulder, I feel like a phone slipping free from its charger. As I close my eyes, adjusting to my new equilibrium, I feel a sense of calm. I know that I will always be a part of the Earth's life force and it will always be a part of me. Our time on this planet is both ephemeral and eternal.

As I absorb the tranquility, deep thoughts upwell in my mind: how can we ever find meaning in a world that is at once both heaven and hell? Is it possible to find joy in a time of great loss and not lose hope in the promise of rebirth?

Coming to terms with the reality of climate change often forces us to grapple with a range of complex emotions. In Caroline Hickman's groundbreaking study on climate change anxiety and depression, researchers reported that close to half of the young people they interviewed said their feelings about climate change negatively affected their daily life and functioning. Three quarters feel the future is frightening, with over half believing that "humanity is doomed." Many fear that the things they value the most will be destroyed. Close to two thirds of young people feel like their government is failing them and betraying future generations. They confessed that significant emotional distress sometimes impacted their ability to function. The top three words used to describe their response were "sad," "afraid" and "anxious," with around 40 percent of people reporting feeling despair, grief or depressed. Less than a third felt optimistic about the future, with 41 percent admitting their reluctance to have children. Perhaps the most confronting finding was that 83 percent of young people around the world believe that people have failed to care for the planet. These sentiments reflect a heartbreaking loss of faith in humanity that threatens to rob us of joy and vitality, undermining our collective response to the crisis we now face.

The implications of Hickman's results are clear: we need to acknowledge that climate change is having far-reaching consequences on our mental health. It's time to start having honest conversations that normalize the strong emotional responses many of us have to the distressing times we are living through. In her book *Facing the Climate Emergency*, American psychologist Margaret Klein Salamon reminds us:

> Your painful feelings spring from the best parts of yourself, from your empathy, sense of responsibility, love for others, and love of life. These feelings connect you to all life and will fuel the work ahead . . . The work you do in learning to accept the feelings that arise in your personal life will enable you to accept, to feel and to use the intense emotional reactions that will result from living your life in climate truth. When you accept and process your painful feelings in your personal life, you are also better able to access, accept and process the pain of others. The more comfortable and confident you are with fear and pain, the more you will be able to help others accept their own

intense feelings and turn them into action. This, too, will be necessary because, as we must strive to remember, we are all on this roller-coaster together.

While some of the stories gathered in beautiful books like Ayana Elizabeth Johnson and Katharine Wilkinson's *All We Can Save* and Cameron Muir, Kirsten Wehner and Jenny Newell's *Living with the Anthropocene* go some way to starting these conversations, we need to be brave enough to speak to the people in our own lives about how we really feel about our changing world. It isn't always easy, but it is essential if we are to try and find our way through this together as a community. Organizations like Psychology for a Safe Climate in Australia, the Climate Psychology Alliance in the United Kingdom, and Climate Awakening in the United States are helping many people navigate difficult psychological terrain. It's vitally important work that I hope helps normalize deeper conversations about the collective challenges we face.

In my own life, people often ask me how I manage to find hope in such a fractured and demoralizing world, particularly given my line of work. My most honest answer is: it isn't always easy. It depends on which day you catch me. I've come to understand that, for a range of complex reasons, some people are just more sensitive than others. We aren't all as thick-skinned as each other. Like far too many of us, I've experienced trauma in my life that makes trusting the inherent goodness in people my biggest challenge. I'm slowly coming to terms with the fact that it's okay to be sensitive, even in my role as a scientist. I'm learning that emotional honesty is something that should be honored and protected, not shunned and attacked, even if the culture of science is still dominated by men who often struggle to articulate their feelings. I hope that sharing my own emotional response to our changing world gives others permission to do the same. Because time is running out, and rigid logic will only get us so far. We can still be rational and professional, while being real and humane. In the words of fellow climate scientist Jeffery Kiehl: "I do not become less of a scientist by opening my heart to the world. I become more whole."

Choosing to be authentic about my own struggle to maintain faith in people during such dark times has not been easy. But it feels important to acknowledge that, at times, our private lives can overpower our ability to participate in collective conversations. I want to be honest about times when I've felt delicate and in need of protection from the brutality I see in the world. Deciding how much of my personal life to share in this book sits somewhere on the spectrum between a generous act of bravery and career suicide. But I've come to understand that many people often feel so disillusioned with others that it stops them from engaging beyond the safety of their own world, let alone even thinking about what they can do to save the planet. So I'm hoping it might help to be a bit more open about my own dilemma with this.

My own world view has been influenced by some very difficult personal experiences, some of which I am still coming to terms with, years and decades on. It's why writing this book has been hard: I'm still struggling to find my own way through. I've come to realize that sadness and joy often coexist; so do hope and grief. I'm still finding the balance in my own life between solitude and connection, the self and community. There are days when the complexities of my inner and outer worlds collide, and depression saturates my every pore:

> Something inside me feels like it has snapped, that some essential thread of hope has failed. Now I find myself in freefall, the horrible familiarity of depression taking hold, all the color and joy draining away from my world. The knowing that sometimes things can't be saved, that the planet is dying, that we couldn't get it together in time to save the irreplaceable. When all the giant trees in a forest finally

disappear. The last of our ancient ancestors lost forever. It feels like so much sadness is upwelling in me right now, like I've tapped into a deep underground chamber of unprocessed grief. It's that helplessness you feel when there is nothing that can be done, that something vital has slipped away. It's hard to know how to keep my head above water at times like this. Maybe the wisest thing I can do right now is just let this grief flow through me.

And sometimes, when things get really bad, I can't imagine ever feeling better:

I feel hollowed out and distant from everything and everyone that I love. I find myself weeping, overwhelmed by feelings of sadness. The waves of tears just keep coming. I can't believe I'm back in this place. The past four months have been such a relief, like coming up for air after nearly drowning. But here I am again, dragged under by that terrible knowing in the pit of my stomach that something has come adrift . . . nothing ever seems to fit together the same way. It feels like every time I go through this, the pain cuts a little deeper. These days, every time I come apart, I shatter into smaller and smaller pieces. I've been broken and reset so many times that I fear that one day things aren't going to be able to come back together again. Every time it happens, I lose a part of myself that I can never regain. It weakens the strength of my heart and my will to live.

I've come to accept that depression is a wound that never heals. It stabilizes from time to time, but it always eventually reopens. The smallest wrong move, and things start to bleed all over again. The more honest I have become with myself and other people about my fragility, the more alienating things can sometimes feel. I've found that, for a range of reasons, many people are afraid to talk about the darker places in themselves. They interpret it as evidence that they are somehow defective, rather than just being a sign of the depth of their human-ity and empathy for our world. Many fear that if they allow others to glimpse the truth of what lies beneath the surface, they will be rejected for being too dark or too complicated. They don't want others to con-front too much pain and reality in their presence. It feels taboo. Or perhaps some people just prefer the shallows. But others, like me, have a tendency towards the deep ocean trenches. It's all okay, just different. People can only ever meet you as deeply as they have met themselves. You just need to find your tribe, and they long to find you.

I know it's easier to try and not feel what you feel, or just hide away from others to keep yourself safe. But after a lifetime of struggling with this, I've made the choice to finally surrender and emotionally inhabit my life. With that, comes the challenge of navigating troubled waters:

> Without warning, I'm plunged deep into dark water, intense pressure is bearing in all around me. I stop resisting and let the stillness and silence saturate me ... I can sink deeper, filling up with more pain, my world buckling and collapsing in around me. Or I can choose to let it go and allow myself to slowly float back to the surface through the blackness, to let the light and air restore me.

I've learned that it is one thing to understand something intellec-tually, and another thing to really feel something in your heart. When losing the thing you love triggers a cascade of pain that leaves you sobbing, you come to understand the true value of the irreplaceable. The agony of this kind of loss is deep, sharp and enduring. It leaves you aching, each day, until every shard either eventually dislodges or embeds even deeper. It's a permeance you must learn to make peace with; it becomes a part of who you are.

When my heart feels tender, I'm learning to let the balm of love and nature gently piece me back together again. Like the golden seams of kintsugi, perhaps our own cracks are the richest parts of ourselves, and mending each other's breaks is a mutual act of healing. People are always going to hurt and disappoint us. That's life; it's the human condition. But if you aren't careful, a few bad experiences can really damage your world view. It's figuring out how to stay connected with people who restore your faith in all that is good, the people who make life worth living. The struggle is to hold onto the belief that there is still joy to be had in life, in the simple, spontaneous moments when you give yourself over to the moment and allow yourself to be moved by the beauty that still surrounds us. When people feel too hard, I look to the natural world. The fern unfurling in the garden, its translucent fronds still untempered by the elements. Holding the gaze of a kookaburra sheltering from the rain as it settles before taking flight. Drifting with the drag of the waves that gently tosses you through the sandy shallows at low tide.

The sense that nothing can or ever will change is one of the hallmarks of depression. As Rebecca Solnit puts it, "the despair felt like a stall, a becalming, a running aground." Believing that things will ever get better is still something I grapple with when things get really bad. Although I've cycled through this enough times to know that these feelings will eventually pass, when I'm in the grip of darkness, I honestly don't believe it ever will. I'm slowly learning that I need to be patient and remember that, eventually, the wind will pick up and catch my sails once more. I've also discovered that I have people in my life who are willing to wade in and give me a nudge, until the stick of the sand beneath me gives way and I'm back in open water.

But when I boil it all down, what gives me hope comes down to this: there is still so much goodness in humanity. Even if you can't see it around you, or have stopped believing that it even exists, there is something good in all of us. When I'm really down, I have to trust that eventually, when the darkness has finally dissipated, I will be able to see again. Until then, I need to hold on and be guided by the light in others. When I allow myself to shelter in their compassion and care, it stills my own flickering light, until eventually, it starts to steady itself and glow a little brighter.

*   *   *

The longer I think about the monumental challenges we face, the more I realize that all any of us can do is choose how we balance our sense of despair and disillusionment with hope and joy in our own lives, every single day. It's not naïve optimism, but a series of pragmatic actions we take to protect our mental health and strengthen our sense of solidarity with other people who also care about our planet. In the environmental classic *Active Hope: How to Face the Mess We're in Without Going Crazy*, Joanna Macy and Chris Johnstone call this "active hope"; something you do rather than something you have. Instead of being passive in the face of life's challenges, active hope is about becoming a proactive agent for the change you want to see in the world. Whether you are a business leader, a student, a musician or a parent, we can all do something – however small it might seem – to influence the cultural evolution of the communities we are a part of. We need to be careful not to fall into the trap of viewing anything less than total victory as a failure; otherwise, it's easy to give up and not celebrate the imperfect wins along the way.

There is so much power in realizing that you can do something to inspire others during these dark times simply by showing up. You can be someone who helps others maintain a belief in the fact that most people are inherently good. While it's true that there are psychopaths, sociopaths and narcissists out there who genuinely don't care about other people because of their own psychological damage, they represent a very small minority of our communities. While we all behave badly sometimes, most people are honestly doing the best that they can from day to day. Most ordinary people, deep down, really do care about the planetary crisis we are facing, but they often feel powerless and disillusioned about their ability to influence change. But the truth is that we are living through a time of history in the making. As Belgian writer Raoul Vaneigem puts it, "revolutionary moments are carnivals in which the individual life celebrates its unification with a regenerated society." We can reclaim our power by being creators of our new world, in the present moment – right now – instead of being fixated on an uncertain future. As Rebecca Solnit writes, hope becomes "an electrifying force in the present" that allows us to participate in inventing and reshaping our vision for a new world.

It is important to remember that if you are reading this book, your capacity to respond is much greater than that of a refugee living in temporary housing or someone from the developing world who struggles to access education and electricity. Instead of being someone who confirms someone's distrust of the basic goodness in humanity, you can choose to be an oasis of hope and sanity in someone else's life. We can allow ourselves to be inspired by others and, when the time comes, choose to be a light in dark places for those around us. As more and more beacons light up around the world, we will illuminate the path forward for others yet to start their own journey home. When we choose to bear witness instead of turning away, we become part of a vital process that will transform our society. In the words of American psychologist and sociologist Melanie Joy:

> Collective witnessing leads to an informed public and a system in which values and practices are more aligned. Think about it: virtually every atrocity in the history of humankind was enabled by a populace that turned away from a reality that seemed too painful to face, while virtually every revolution for peace and justice has been made possible by a group of people who chose to bear witness and demanded

that others bear witness as well. The goal of all justice movements is to activate collective witnessing so that social practices reflect social values. A movement succeeds when it reaches a critical mass of witnesses – that is, enough witnesses to tip the scales of power in favor of the movement.

As we live through this time of great upheaval, we must also hold a place in our hearts for healing; healing within ourselves, the people we love, and even those who hurt us and the world. We can choose which side of history we want to be on and make the personal choices that help make the world a better place. So many of us have lost faith in the goodness in humanity, we've lost touch with our inner knowing of what is true and whole in ourselves and each other. When we contact this universal place of compassion and the interconnectedness of our hearts, we experience a sense of homecoming, a deeply felt sense of our belonging to our shared humanity.

Perhaps we need to start telling ourselves different stories, ones that remind us of the best part of ourselves. In her brilliant books *A Paradise Built in Hell* and *Hope in the Dark*, Rebecca Solnit writes of the extraordinary community spirit that arises in the face of disasters. She writes of the altruism, resourcefulness and generosity that seem to define people facing unthinkable challenges. In the aftermath of Hurricane Katrina, a devastating category 5 system that struck the United States in 2005, the true beauty of humanity shone through:

Hundreds of boat owners rescued people – single moms, toddlers, grandfathers – stranded in attics, on roofs, in flooded housing projects, hospitals and school buildings. None of them said, I can't rescue everyone, therefore it is futile; therefore my efforts are flawed and worthless, though that's often what people say about more abstract issues in which, nevertheless, lives, places, cultures, species, rights are at stake ... None of those people said I can't rescue them all.

All of them said, I can rescue someone, and that's work so meaning-ful and important I will risk my life and defy the authorities to do it. And they did.

In the final stages of working on this book, as if on cue, one of the most catastrophic flood events in Australian history engulfed the east coast of the country. It was surreal to witness scenes of complete devas-tation along the eastern seaboard where I live, literally the same day the IPCC's *Sixth Assessment Report on Impacts, Adaptation and Vulnerabil-ity* - the "atlas of human suffering" - was released on 28 February 2022. Historical rainfall records were broken in a handful of days, resulting in the unprecedented flooding of my husband's hometown of Lismore in northern New South Wales, around 50 kilometers from where we live. It drove home in the most personal way that climate change isn't just about numbers on a graph - it is about the people and places we love. It's about the stability of life as we know it. How local communities choose to respond to climate change is literally a matter of life or death.

As one of the lead authors on the IPCC's chapter on global water cycle changes, I am well aware that heavy rainfall events are expected to intensify as the planet continues to warm, increasing flood risk. Following a wet spring and summer caused by La Niña conditions in the Pacific Ocean, the catchments of eastern Australia were satu-rated and primed for flooding. At Dunoon, just north of Lismore, 775 millimeters of rain fell in twenty-four hours - more than the cities of London or Melbourne receive in an entire year. Before February 2022, the highest flood levels in Lismore were recorded in 1954 and 1974, when flood heights reached a maximum of 12.27 meters and 12.15 meters respectively. When you walk around town, the 1974 flood heights tower above you on power poles in the central business dis-trict, a reminder of a dramatic past. But the 2022 flood was not caused by cyclonic conditions or preceded by the prolonged wet conditions that characterized much of the 1950s and 1970s. Instead, the region was coming off the back of a severe drought that lasted from 2017 until 2020. After Australia's hottest and driest year on record – the landscape was so bone-dry that even rainforests burnt during our Black Summer – I thought it would take years before it would become saturated enough to seriously flood again.

When the flood warnings were issued for the Wilsons River that flows through Lismore, authorities predicted the river would rise 11.50 meters, similar to the heights reached in the 2017 floods caused by ex-Tropical Cyclone Debbie. Locals in flood-prone areas responded, moving as much as possible to higher ground. People worked for hours, filling the upper storys, attics and mezzanines of houses and businesses to above the 1974 flood height, which has long been used as a bench-mark to prepare for floods. That night, my husband called his sister, whose home was above historical flood levels, to move her car to higher ground. Just in case. I'd been watching the torrential rain battering the city of Brisbane, around 200 kilometers to the north, so I had an uneasy feeling about the extreme conditions starting to drift farther south. The last thing I texted her before going to bed that night was: "Brisbane has now broken its 1974 rainfall record, expect the unexpected."

By three a.m. next morning, the Wilsons River had broken its banks and although my sister-in-law's home was on stilts on the very edge of the 1974 flood zone, water had started filling her backyard and was rising fast up the back stairs. They rushed to stash their most valuable possessions higher – on makeshift platforms made by taking doors off their hinges and placing them on top of high-standing furniture – but it was soon clear they couldn't save much with the time they had left. As an artist, my sister-in-law's priority was to save key works from her latest collection – ironically, a series of climate change–themed works celebrating Australia's unique wildlife. Her entire studio, containing over twenty years of work, would soon be underwater.

Within an hour, the water had come up so high that they had to leave immediately or risk getting trapped inside. She wrapped her terrified cat in a blanket and put it in a plastic laundry tub, along with a mobile phone, a charger and a toothbrush, and swam off the front verandah of her home into muddy floodwater in the predawn darkness. Her feet couldn't touch the bottom until she swam around 100 meters up her street. By sun-up, floodwaters lapped at the roof of her home. She was alive, but now had nothing but the clothes on her back.

While it's easy to cast judgment on people who live in flood-prone towns like Lismore, it is unfair to claim they were unprepared. It's impossible to prepare for a flood that would go on to obliterate historical flood levels by a full 2 meters. I tracked the Wilsons River flood level online, watching in horror as it reached a staggering 14.40 meters. Climate records are not usually broken by such monstrous margins. I couldn't comprehend the implications of what I was witnessing; it was sickening.

To make matters worse, the army did not show up until a full five days later, leaving the community to fend for themselves. Thankfully the people of Lismore are one of the most resilient, flood-prepared communities in Australia. We saw extraordinary scenes of locals rescuing each other from rooftops in their boats, jet skis and kayaks. They didn't wait for the handful of official rescue boats to be deployed. Our family relayed stories of people in more remote areas setting off in their boats with cordless angle grinders to cut people out of their homes' roof cavities, where they'd been forced to retreat. Trying to sleep with the sound of landslides crashing through the valleys. Days of fuel and food shortages. The mud. The terrible stench of death and debris. A town utterly destroyed.

Thousands of homes were badly damaged or destroyed beyond repair, leaving many people displaced or homeless. Many have lost everything and can't afford to rebuild. Members of my own family have lost their homes, businesses and/or all of their possessions to this disaster. As I write this, two weeks after the floodwaters receded, I still have family trapped by landslides without electricity, as power-lines have tumbled down saturated hillslopes that collapsed under the weight of torrential rain. My father-in-law's business of close to thirty years was completely destroyed, roller doors bowed by the weight of floodwaters and the contents of a motorcycle mechanic's workshop churned like the spin cycle of a colossal washing machine. When asked if the army had been past to help, he replied, "What army?"

The only army that turned up for most people was a ragtag col-lection of friends, family and volunteers doing their best to help their community in the face of unimaginable destruction. I was moved by the stories and footage that emerged in the aftermath of the event. Locals evacuated from their own homes returning to rescue others. People from surrounding towns flocking in to help overwhelmed locals remove the entire contents of their homes and businesses into gigantic rubbish mounds to then begin the slog of sweeping river mud and debris from the remnants of their former lives. The crowdfunding of private helicop-ters to drop off supplies to people in remote communities cut off by the floodwaters. The group of Fijian men, workers from the local abattoir, singing traditional songs from their island home to soothe traumatized locals gathered in evacuation shelters. Who better to understand the profound loss of displacement than the people of the Pacific? Witness-ing their compassion opened the floodgate to my tears – for my family, for my country. For our world. When biblical floods descended on our nation, it wasn't the government that came to the rescue; it was the local community that showed up. Without their care and bravery, countless more people would have died. Without our humanity, we are no one.

The most disturbing thing for me as a scientist is that Australia's east coast floods were completely unprecedented, totally unpredicted by the models. They defied logic, pushing us far from the safe shores of our recorded history. And if this is how underprepared a wealthy country like Australia is for escalating climate extremes, then imagine the situation that will unfold across the developing world.

This madness must stop. The era of fossil fuels must come to an end.

* * *

As Mahatma Gandhi once said, "the greatness of humanity is not in being human, but in being humane." Only when we return to a place of shared compassion can we truly begin the work of restoring hope and life on our planet. When we reawaken our senses that have become numbed and estranged from the world around us, we experience a profound sense of our belonging in this extraordinary web of life, a reverence for our miraculous planet, and an awareness of the power we all wield to harm or heal, to accept or abandon our unique role in creating a more inspiring future. When we remember that each of us has a part to play, when we believe that our contribution really does matter, our soul comes alive. We find meaning in a meaningless world.

Because ultimately, it's having enough people who care, who are willing to re-emerge from our collective amnesia and correct our course, to create a better future based on the foundation of hope, healing and unity. We must ask ourselves whether we want to be part of strengthening the underlying fabric of society or be complicit in the erosion of the shared values that bind all of humanity. When we reawaken that part of ourselves that believes in change and healing, we inspire those still stumbling, terrified and hopeless in the dark, to consider that another reality is possible. We all have our own inner demons to conquer, but not enough of us realize that much of what lurks in the shadows is healed in our reconnection with ourselves, each other and the natural world. When we realize the power we have to be a part of someone else's healing, it becomes effortless to extend this care to the rest of the world and help restore life on our planet. You just need to show up for the opportunities that present themselves in your life.

The time has come to repair our tragic severing from the web of belonging all around us. Only then will true change really happen, when our individual transformations unite to evolve the collective cultures that shape our societies. On my good days I feel a true sense of this possibility: that when we open our hearts and return to each other, we will eventually heal our world. Because ultimately, we have a choice to make about whether we turn towards or away from each other to meet the shared challenges we face. Although we all go through times when we feel overwhelmed and demoralized by what we see all around us, it's important to remember that the truest part of ourself wants to restore our relationship between our inner and outer worlds. It's part of our evolutionary wiring: we are relational beings, we all long to belong to something bigger than ourselves. We are all part of the Earth's life force.

As someone who manages depression, I understand that sometimes all we can do is try and keep our head above water, taking each day one step at a time even when we are ragged and weary with exhaustion and heartache. During those times, remember to be kind to yourself and know that it's normal to feel a range of emotions. Although it sometimes doesn't feel like it, eventually things start to shift, and the pain will eventually pass, one way or another. As many of us know, a simple act of kindness directed towards us when we feel raw and fragile can restore our faith in people.

When we choose to open ourselves to our vulnerability and our pain, we realize that it is not just our individual sorrow that we feel; our suffering is also part of a broader, collective suffering that comes from our severed belonging from ourselves and each other. When we abandon our values and shrink into the tight, regressed places within ourselves, we choose scarcity over abundance, exile over homecoming. When our own light has gone out, we must trust that we can be led by the light that shines in others, that we are part of a web of goodwill that surrounds each of us and connects the world. In these moments we have to trust in the goodness of others until the storms in our internal world have passed over and we can see clearly once more.

Because we have all been born into unique circumstances of family, culture, opportunity, gender and geography, we all have something uniquely our own to offer the world. Like each voice in a choir, our contribution to life's music becomes part of a force that enriches us all. When we uncover what it is that yearns to be expressed through us, we transcend the confines of our personal circumstances and contribute to the power of the collective. We tap into the eternal force that has transformed humanity throughout our history. If we are brave enough to face the disconnect and loneliness in ourselves, we discover a universal sense of alienation and despair that has engulfed much of humanity. As we have been driven away from each other and the natural world by technology and mass consumerism, we now face a primal sense of estrangement that can only be healed by reawakening our senses and allowing ourselves to be moved by what we see around us. When we allow ourselves to acknowledge our isolation, pain and disconnection from ourselves, each other and the natural world, it will lead us back to universal truths that govern humanity: we are all animals that have a finite time left on this beautiful, battered Earth of ours. What we each do with our one wild and precious life ripples out and affects us all.

As a university lecturer, I know many young people feel afflicted by the challenges we face; students wonder why they have to be the ones to live through an era of so much destruction. Although I don't really have an answer to that question, I am comforted by the words of J.R.R. Tolkien in his epic fantasy tale *The Fellowship of the Ring*, when Frodo, greatly outnumbered and battle-weary, laments to the wise wizard Gandalf the grave burden he feels to save Middle Earth from the dark forces of Sauron:

"I wish it need not have happened in my time," said Frodo.

"So do I," said Gandalf, "and so do all who live to see such times. But that is not for them to decide. All we have to decide is what to do with the time that is given us."

It's also helpful to remember that we are not the first to live through difficult times. The struggle for justice has been with us throughout the ages. It's a fundamental part of the human condition. Each generation contributes to humanity's deeply layered history. Perhaps we are all simply here to help make the world a better place in any way we can, however big or small. We can choose to be a force of goodness and healing, over disharmony and destruction. That choice is ours, every single day, in our relationships with each other and the way that we acknowledge our place in our shared web of life. We can choose to turn a blind eye, or be moved by what we see, allow it to be a force that compels us forward and elevates us into being the best person that we can be. We can choose whether we answer the call of our own personal growth and the collective evolution of our species. As David Attenborough writes in *A Life on Our Planet*:

> We have come as far as we have because we are the cleverest creatures to have ever lived on Earth. But if we are to continue to exist, we will require more than intelligence. We will require wisdom.

We can choose to answer the call of history with our head, our heart and as part of our communities, and steer ourselves towards a brighter future. It promises to be humanity's finest moment. The moment when we realize that we are each part of an eternal, evolutionary force for change; a single raindrop that helps form a vast ocean. When we awaken this miraculous potential that exists within us all, we can and will change our world.

Acknowledgments

THIS BOOK WAS WRITTEN ON the lands of the Minjungbal people of the Bundjalung nation in northern New South Wales, Australia. Pure medicine country. It is a place of outstanding natural beauty that countless generations have fought hard to protect. I am privileged to live in a community where a strong custodian spirit for the natural world still thrives. Its wisdom and beauty is infused in these pages.

This book would not have been possible without the support of my agent, Jane Novak; thank you for being such a strong champion of my writing and for steering me out of rough seas with your formidable blend of pragmatism and care. I would have been lost out there without you.

I am very grateful to my team at Black Inc. for making this book something worth sharing far and wide: Sophy Williams, Julia Carlomagno, Denise O'Dea, Jo Rosenberg, Kate Nash, Erin Sandiford and Nina Kenwood. Special thanks to Julia for encouraging me to write a book to "help fuel the social movement," to Sophy for overseeing the unforeseen challenges along the way with empathy and respect, and to Denise for your editorial light touch. Thanks to each of you for being passionate advocates for this work; your contributions helped bring this story to life and find its audience. Thanks also to Akiko Chan for incorporating the beauty of Australia's native plants on the cover art for me.

Many thanks to Sian Prior, firstly for pointing me in Jane's direction, but most of all for instilling confidence in me as a writer. You taught me how to balance sensitivity and intelligence in my work. Thank you for being an inspiration for women like me. I want to thank my Melbourne writing buddies, Julie Perrin, Annie Keely and Michael Green, for the messages, chats, flowers and cards at times when I needed reminding that what I have to say really matters. Your friendship helped me believe that kind, switched-on people are actually listening and really do care. Thank you for your ongoing love and belief in me. I am also grateful to others from the writing and publishing world: Ashley Hay, Nick Feik, Sophie Cunningham, Bridie Jabour, Nicole Hasham, Eugenie Kelly and Erik Jensen for providing a scientist like me with an audience. This book is a testament to your encouragement and support.

I want to acknowledge the hundreds of IPCC scientists and Technical Support Unit staff who contributed their blood, sweat and tears to the delivery of the *Sixth Assessment Report*. It was the ultimate challenge, made even harder by a deadly pandemic and a string of escalating extremes. So many of you went above and beyond the call of duty. It took its toll. Know that future generations will be grateful that we did all that we could. I am honored to have worked with such an exceptionally dedicated group of altruistic people. IPCC reports alone are proof of the existence of the good in humanity. I hope I have done our work justice in these pages.

I am particularly grateful to my colleagues who helped make this book possible. Special thanks to Professor James Renwick for such a careful and thorough review of the manuscript; your astute technical comments and words of encouragement meant the world to me. Thank you for your camaraderie through the IPCC process and your effort helping me refine this work – I am very grateful for your generous support. Thanks also to Professor David Karoly, my long-time mentor and friend, for reviewing the manuscript through the eyes of an IPCC veteran who has contributed to every single IPCC assessment report in some way. Your dedication to our field is the stuff of legends. You have inspired me, and countless others, to pursue this path. Thank you for all of your guidance and support throughout my career, and for encouraging me to be the writer that I am. Many thanks to Zak Baillie for keeping my research projects alive while I was juggling IPCC and teaching, and again as I diverted all of my energy into writing this book. Thank you for all of your technical support and good-natured companionship through turbulent times. Working with you gives me real hope in the younger generations. Particular thanks to Saul Cunningham, for being an inspiring leader and for providing ongoing support during a professionally and personally challenging period. Many thanks to Gail Frank for your good-humoured guidance on managing the pressures of a demanding career.

My deepest thanks go to my inner circle, the people who love me for who I am, not what I do: Josh, Kimberley, Madeleine, Katy, Bec, Emma, Stefanie and Dave. I am honoured to walk this path with you. Thank you for helping me understand that you are never a burden on people who truly love you. Special thanks to Stef for reading this manuscript through the sensitive eyes of an artist – your perspective was a gift; to Madeleine, for staying with me, week after week, year after year, as I've grappled with my darkness – none of this would be possible without your care and compassion; and to Kimberley, my goddess, for showing me the true meaning of unconditional love. You have taught me that everything is welcome, and there is always a way back home. And finally, Josh; my heaven on Earth. Thank you for the blessings of a harmonious life; you are my greatest inspiration, wisest counsel and most precious friend. My love for you is eternal.

References

PART 1

Abram, N. J., Henley, B. J., Sen Gupta, A., Lippmann, T. J. R., Clarke, H., Dowdy, A. J., Sharples, J. J., Nolan, R. H., Zhang, T., Wooster, M. J., Wurtzel, J. B., Meissner, K. J., Pitman, A. J., Ukkola, A. M., Murphy, B. P., Tapper, N. J. and Boer, M. M. (2021). Connections of climate change and variability to large and extreme forest fires in southeast Australia. *Communications Earth & Environment* 2 (1): https://doi.org/10.1038/s43247-020-00065-8.

Arctic Monitoring and Assessment Programme (2021). *Arctic Climate Change Update 2021: Key Trends and Impacts. Summary for Policymakers.* Arctic Monitoring and Assessment Programme (AMAP), Tromsø, Norway. Accessed at: https://www.amap.no/documents/doc/arctic-climate-change-update-2021-key-trends-and-impacts.-summary-for-policy-makers/3508.

Attenborough, D. (2020). *A Life on Our Planet: My Witness Statement and Vision for the Future.* Penguin Random House, London, UK.

Bastin, J.-F., Clark, E., Elliott, T., Hart, S., van den Hoogen, J., Hordijk, I., Ma, H., Majumder, S., Manoli, G., Maschler, J., Mo, L., Routh, D., Yu, K., Zohner, C. M. and Crowther, T. W. (2019). Understanding climate change from a global analysis of city analogues. *PLOS ONE* 14 (7): e0217592.

Boer, M. M., Resco de Dios, V. and Bradstock, R. A. (2020). Unprecedented burn area of Australian mega forest fires. *Nature Climate Change* 10: 171-172.

Brovkin, V., Brook, E., Williams, J. W., Bathiany, S., Lenton, T. M., Barton, M., DeConto, R. M., Donges, J. F., Ganopolski, A., McManus, J., Praetorius, S., de Vernal, A., Abe-Ouchi, A., Cheng, H., Claussen, M., Crucifix, M., Gallopín, G., Iglesias, V., Kaufman, D. S., Kleinen, T., Lambert, F., van der Leeuw, S., Liddy, H., Loutre, M.-F., McGee, D., Rehfeld, K., Rhodes, R., Seddon, A. W. R., Trauth, M. H., Vanderveken, L. and Yu, Z. (2021). Past abrupt changes, tipping points and cascading impacts in the Earth system. *Nature Geoscience* 14: 550-558.

Caesar, L., McCarthy, G. D., Thornalley, D. J. R., Cahill, N. and Rahmstorf, S. (2021). Current Atlantic Meridional Overturning Circulation weakest in last millennium. *Nature Geoscience* 14: 118-120.

Caesar, L., Rahmstorf, S., Robinson, A., Feulner, G. and Saba, V. (2018). Observed fingerprint of a weakening Atlantic Ocean Overturning Circulation. *Nature* 556 (7700): 191–196.

Carbon Brief (2021). Do COP26 promises keep global warming below 2C? Accessed at: https://www.carbonbrief.org/analysis-do-cop26-promises-keep-global-warming-below-2c.

Carson, R. (1962). *Silent Spring.* Haughton Mifflin, Boston, USA.

Cooper, G. S., Willcock, S. and Dearing, J. A. (2020). Regime shifts occur disproportionately faster in larger ecosystems. *Nature Communications* 11 (1): 1175.

Drijfhout, S., Bathiany, S., Beaulieu, C., Brovkin, V., Claussen, M., Huntingford, C., Scheffer, M., Sgubin, G. and Swingedouw, D. (2015). Catalogue of abrupt shifts in Intergovernmental Panel on Climate Change climate models. *Proceedings of the National Academy of Sciences* 112 (43): E5777.

Dutton, A., Carlson, A. E., Long, A. J., Milne, G. A., Clark, P. U., DeConto, R., Horton, B. P., Rahmstorf, S. and Raymo, M. E. (2015). Sea-level rise due to polar ice-sheet mass loss during past warm periods. *Science* 349 (6244): DOI 10.1126/science.aaa4019.

Energy and Climate Intelligence Unit (2022). Net Zero Scorecard 2022. Accessed at: https://eciu.net/netzerotracker.

Feng, X., Merow, C., Liu, Z., Park, D. S., Roehrdanz, P. R., Maitner, B., Newman, E. A., Boyle, B. L., Lien, A., Burger, J. R., Pires, M. M., Brando, P. M., Bush, M. B., McMichael, C. N. H., Neves, D. M., Nikolopoulos, E. I., Saleska, S. R., Hannah, L., Breshears, D. D., Evans, T. P., Soto, J. R., Ernst, K. C. and Enquist, B. J. (2021). How deregulation, drought and increasing fire impact Amazonian biodiversity. *Nature* 597: 516–521.

Fischer, H., Meissner, K. J., Mix, A. C., Abram, N. J., Austermann, J., Brovkin, V., Capron, E., Colombaroli, D., Daniau, A.-L., Dyez, K. A., Felis, T., Finkelstein, S. A., Jaccard, S. L., McClymont, E. L., Rovere, A., Sutter, J., Wolff, E. W., Affolter, S., Bakker, P., Ballesteros-Cánovas, J. A., Barbante, C., Caley, T., Carlson, A. E., Churakova, O., Cortese, G., Cumming, B. F., Davis, B. A. S., de Vernal, A., Emile-Geay, J., Fritz, S. C., Gierz, P., Gottschalk, J., Holloway, M. D., Joos, F., Kucera, M., Loutre, M.-F., Lunt, D. J., Marcisz, K., Marlon, J. R., Martinez, P., Masson-Delmotte, V., Nehrbass-Ahles, C., Otto-Bliesner, B. L., Raible, C. C., Risebrobakken, B., Sánchez Goñi, M. F., Arrigo, J. S., Sarnthein, M., Sjolte, J., Stocker, T. F., Velasquez Alvárez, P. A., Tinner, W., Valdes, P. J., Vogel, H., Wanner, H., Yan, Q., Yu, Z., Ziegler, M. and Zhou, L. (2018). Palaeoclimate constraints on the impact of 2°C anthropogenic warming and beyond. *Nature Geoscience* 11 (7): 474–485.

Gatti, L. V., Basso, L. S., Miller, J. B., Gloor, M., Gatti Domingues, L., Cassol, H. L. G., Tejada, G., Aragão, L. E. O. C., Nobre, C., Peters, W., Marani, L., Arai, E., Sanches, A. H., Corrêa, S. M., Anderson, L., Von Randow, C., Correia,

C. S. C., Crispim, S. P. and Neves, R. A. L. (2021). Amazonia as a carbon source linked to deforestation and climate change. *Nature* 595 (7867): 388–393.

Geyer, R., Jambeck Jenna, R. and Law Kara, L. Production, use, and fate of all plastics ever made. *Science Advances* 3 (7): e1700782.

Grose, M., Narsey, S., Delage, F.P., Dowdy, A.J., Bador, M., Boschat, G., Chung, C., Kajtar, J.B., Rauniyar, S., Freund, M.B., Lyu, K., Rashid, H., Zhang, X., Wales, S., Trenham, C., Holbrook, N.J., Cowan, T., Alexander, L., Arblaster, J.M. and Power, S. (2020). Insights from CMIP6 for Australia's future climate. *Earth's Future* 8 (e2019EF001469): 1–24.

Hansen, J. E. (2007). Scientific reticence and sea level rise. *Environmental Research Letters* 2 (2): 024002.

Hooijer, A. and Vernimmen, R. (2021). Global LiDAR land elevation data reveal greatest sea-level rise vulnerability in the tropics. *Nature Communications* 12 (1): 3592.

International Cryosphere Climate Initiative (2021). *State of the Cryosphere 2021: A Needed Decade of Urgent Action.* Accessed at: http://iccinet.org/statecryo21/.

IPCC (2021). Summary for Policymakers. In *Climate Change 2021: The Physical Science Basis. Contribution of Working Group I to the Sixth Assessment Report of the Intergovernmental Panel on Climate Change.* Masson-Delmotte, V., Zhai, P., Pirani, A., Connors, S. L., Péan, C., Berger, S., Caud, N., Chen, Y., Goldfarb, L., Gomis, M. I., Huang, M., Leitzell, K., Lonnoy, E., Matthews, J. B. R., Maycock, T. K., Waterfield, T., Yelekçi, O., Yu, R. and Zhou, B. (eds). Accessed at: https://www.ipcc.ch/report/ar6/wg1/.

IPCC (2022). Summary for Policymakers. In *Climate Change 2022: Mitigation of Climate Change, the Working Group III contribution to the Sixth Assessment Report of the Intergovernmental Panel on Climate Change.* Masson-Delmotte, V., Zhai, P., Pirani, A., Connors, S. L., Péan, C., Berger, S., Caud, N., Chen, Y., Goldfarb, L., Gomis, M. I., Huang, M., Leitzell, K., Lonnoy, E., Matthews, J. B. R., Maycock, T. K., Waterfield, T., Yelekçi, O., Yu, R. and Zhou, B. (eds). Accessed at: https://www.ipcc.ch/report/ar6/wg3/.

IPCC, Chen, D., Rojas, M., Samset, B. H., Cobb, K., Diongue Niang, A., Edwards, P., Emori, S., Faria, S. H., Hawkins, E., Hope, P., Huybrechts, P., Meinshausen, M., Mustafa, S. K., Plattner, G. K. and Tréguier, A. M. (2021a). Chapter 1: Framing, Context, and Methods. In *Climate Change 2021: The Physical Science Basis. Contribution of Working Group I to the Sixth Assessment Report of the Intergovernmental Panel on Climate Change.* Masson-Delmotte, V., Zhai, P., Pirani, A., Connors, S. L., Péan, C., Berger, S., Caud, N., Chen, Y., Goldfarb, L., Gomis, M. I., Huang, M., Leitzell, K., Lonnoy, E., Matthews, J. B. R., Maycock, T. K., Waterfield, T., Yelekçi, O., Yu, R. and Zhou, B. (eds). Accessed at: https://www.ipcc.ch/report/ar6/wg1/.

IPCC, Douville, H., Raghavan, K., Renwick, J., Allan, R. P., Arias, P. A., Barlow, M., Cerezo-Mota, R., Cherchi, A., Gan, T. Y., Gergis, J., Jiang, D., A. Khan, A., Pokam Mba, W., Rosenfeld, D., J. Tierney, J. and Zolina, O. (2021b). Chapter 8: Water Cycle Changes. In *Climate Change 2021: The Physical Science Basis. Contribution of Working Group I to the Sixth Assessment Report of the Intergovernmental Panel on Climate Change.* Masson-Delmotte, V., Zhai, P., Pirani, A., Connors, S. L., Péan, C., Berger, S., Caud, N., Chen, Y., Goldfarb, L., Gomis, M. I., Huang, M., Leitzell, K., Lonnoy, E., Matthews, J. B. R., Maycock, T. K., Waterfield, T., Yelekçi, O., Yu, R. and Zhou, B. (eds). Accessed at: https://www.ipcc.ch/report/ar6/wg1/.

IPCC, Forster, P., Storelvmo, T., Armour, K., Collins, W., Dufresne, J. L., Frame, D., Lunt, D. J., Mauritsen, T., Palmer, M. D., Watanabe, M., Wild, M. and Zhang, H. (2021c). Chapte 7: The Earth's Energy Budget, Climate Feedbacks, and Climate Sensitivity. In *Climate Change 2021: The Physical Science Basis. Contribution of Working Group I to the Sixth Assessment Report of the Intergovernmental Panel on Climate Change.* Masson-Delmotte, V., Zhai, P., Pirani, A., Connors, S. L., Péan, C., Berger, S., Caud, N., Chen, Y., Goldfarb, L., Gomis, M. I., Huang, M., Leitzell, K., Lonnoy, E., Matthews, J. B. R., Maycock, T. K., Waterfield, T., Yelekçi, O., Yu, R. and Zhou, B. (eds). Accessed at: https://www.ipcc.ch/report/ar6/wg1/.

IPCC, Fox-Kemper, B., Hewitt, H. T., Xiao, C., Aðalgeirsdóttir, G., Drijfhout, S. S., Edwards, T. L., Golledge, N. R., Hemer, M., Kopp, R. E., Krinner, G., Mix, A., Notz, D., Nowicki, S., Nurhati, I. S., L. Ruiz, L., Sallée, J. B., Slangen, A. B. A. and Yu, Y. (2021d). Chapter 9: Ocean, Cryosphere and Sea Level Change. In *Climate Change 2021: The Physical Science Basis. Contribution of Working Group I to the Sixth Assessment Report of the Intergovernmental Panel on Climate Change.* Masson-Delmotte, V., Zhai, P., Pirani, A., Connors, S. L., Péan, C., Berger, S., Caud, N., Chen, Y., Goldfarb, L., Gomis, M. I., Huang, M., Leitzell, K., Lonnoy, E., Matthews, J. B. R., Maycock, T. K., Waterfield, T., Yelekçi, O., Yu, R. and Zhou, B. (eds). Accessed at: https://www.ipcc.ch/report/ar6/wg1/.

IPCC, Ranasinghe, R., Ruane, A. C., Vautard, R., Arnell, N., Coppola, E., Cruz, F. A., Dessai, S., Islam, A. S., Rahimi, M., Ruiz Carrascal, D., Sillmann, J., Sylla, M. B., Tebaldi, C., Wang, W. and Zaaboul, R. (2021e). Chapter 12: Climate Change Information for Regional Impact and for Risk Assessment. In *Climate Change 2021: The Physical Science Basis. Contribution of Working Group I to the Sixth Assessment Report of the Intergovernmental Panel on Climate Change.* Masson-Delmotte, V., Zhai, P., Pirani, A., Connors, S. L., Péan, C., Berger, S., Caud, N., Chen, Y., Goldfarb, L., Gomis, M. I., Huang, M., Leitzell, K., Lonnoy, E., Matthews, J. B. R., Maycock, T. K., Waterfield, T., Yelekçi, O., Yu, R. and Zhou, B. (eds). Accessed at: https://www.ipcc.ch/report/ar6/wg1/.

IPCC, Seneviratne, S. I., Zhang, X., Adnan, M., Badi, W., Dereczynski, C., Di Luca, A., Ghosh, S., Iskandar, I., Kossin, J., Lewis, S., Otto, F., Pinto, I., Satoh, M., Vicente-Serrano, S. M., Wehner, M. and Zhou, B. (2021f). Chapter 11: Weather and Climate Extreme Events in a Changing Climate. In *Climate Change 2021: The Physical Science Basis. Contribution of Working Group I to the Sixth Assessment Report of the Intergovernmental Panel on Climate Change*. Masson-Delmotte, V., Zhai, P., Pirani, A., Connors, S. L., Péan, C., Berger, S., Caud, N., Chen, Y., Goldfarb, L., Gomis, M. I., Huang, M., Leitzell, K., Lonnoy, E., Matthews, J. B. R., Maycock, T. K., Waterfield, T., Yelekçi, O., Yu, R. and Zhou, B. (eds). Accessed at: https://www.ipcc.ch/report/ar6/wg1/.

IPCC, Canadell, J. G., Monteiro, P. M. S., Costa, M. H., Cotrim da Cunha, L., Cox, P. M., Eliseev, A. V., Henson, S., Ishii, M., Jaccard, S., Koven, C., Lohila, A., Patra, P. K., Piao, S., Rogelj, J., Syampungani, S., Zaehle, S. and Zickfeld, K. (2021). Chapter 5: Global Carbon and other Biogeochemical Cycles and Feedbacks Supplementary Material. In *Climate Change 2021: The Physical Science Basis. Contribution of Working Group I to the Sixth Assessment Report of the Intergovernmental Panel on Climate Change*. Masson-Delmotte, V., Zhai, P., Pirani, A., Connors, S. L., Péan, C., Berger, S., Caud, N., Chen, Y., Goldfarb, L., Gomis, M. I., Huang, M., Leitzell, K., Lonnoy, E., Matthews, J. B. R., Maycock, T. K., Waterfield, T., Yelekçi, O., Yu, R. and Zhou, B. (eds). Accessed at: https://www.ipcc.ch/report/ar6/wg1/.

IPCC, Gutiérrez, J. M., Jones, R. G., Narisma, G. T., Alves, L. M., Amjad, M., Gorodetskaya, I. V., Grose, M., Klutse, N. A. B., Krakovska, S., Li, J., Martínez-Castro, D., Mearns, L. O., Mernild, S. H., Ngo-Duc, T. van den Hurk, B. and Yoon, J.-H. (2021). Atlas. In *Climate Change 2021: The Physical Science Basis. Contribution of Working Group I to the Sixth Assessment Report of the Intergovernmental Panel on Climate Change*. Masson-Delmotte, V., Zhai, P., Pirani, A., Connors, S. L., Péan, C., Berger, S., Caud, N., Chen, Y., Goldfarb, L., Gomis, M. I., Huang, M., Leitzell, K., Lonnoy, E., Matthews, J. B. R., Maycock, T. K., Waterfield, T., Yelekçi, O., Yu, R. and Zhou, B. (eds). Accessed at: https://www.ipcc.ch/report/ar6/wg1/: Available from http://interactive-atlas.ipcc.ch/.

IPCC (2021). Technical Summary. In *Climate Change 2021: The Physical Science Basis. Contribution of Working Group I to the Sixth Assessment Report of the Intergovernmental Panel on Climate Change*. Masson-Delmotte, V., Zhai, P., Pirani, A., Connors, S. L., Péan, C., Berger, S., Caud, N., Chen, Y., Goldfarb, L., Gomis, M. I., Huang, M., Leitzell, K., Lonnoy, E., Matthews, J. B. R., Maycock, T. K., Waterfield, T., Yelekçi, O., Yu, R. and Zhou, B. (eds). Accessed at: https://www.ipcc.ch/report/ar6/wg1/.

Kaufman, D., McKay, N., Routson, C., Erb, M., Dätwyler, C., Sommer, P. S., Heiri, O. and Davis, B. (2020). Holocene global mean surface temperature, a multi-method reconstruction approach. *Scientific Data* 7 (1): 201.

Lade, S. J., Steffen, W., de Vries, W., Carpenter, S. R., Donges, J. F., Gerten, D., Hoff, H., Newbold, T., Richardson, K. and Rockström, J. (2020). Human impacts on planetary boundaries amplified by Earth system interactions. *Nature Sustainability* 3 (2): 119–128.

Lear, C. H., Anand, P., Blenkinsop, T., Foster, G. L., Gagen, M., Hoogakker, B., Larter, R. D., Lunt, D. J., McCave, I. N., McClymont, E., Pancost, R. D., Rickaby, R. E. M., Schultz, D. M., Summerhayes, C., Williams, C. J. R. and Zalasiewicz, J. (2021). Geological Society of London scientific statement: what the geological record tells us about our present and future climate. *Journal of the Geological Society* 178 (1): jgs2020–239.

Lenton, T. M., Rockström, J., Gaffney, O., Rahmstorf, S., Richardson, K., Steffen, W. and Schellnhuber, H. J. (2019). Climate tipping points – too risky to bet against. *Nature* 575: 592–595.

Lloyd, E. A., Oreskes, N., Seneviratne, S. I. and Larson, E. J. (2021). Climate scientists set the bar of proof too high. *Climatic Change* 165: https://doi. org/10.1007/s10584-021-03061-9.

Lovejoy, T. E. and Nobre, C. (2018). Amazon tipping point: last chace for action. *Science Advances* 4 (2): eaat2340.

Meehl, G. A., Senior, C. A., Eyring, V., Flato, G., Lamarque, J.-F., Stouffer, R. J., Taylor, K. E. and Schlund, M. (2020). Context for interpreting equilibrium climate sensitivity and transient climate response from the CMIP6 Earth system models. *Science Advances* 6 (26): DOI 10.1126/sciadv.aba1981.

Meinshausen, M., Lewis, J., McGlade, C., Gütschow, J., Nicholls, Z., Burdon, R., Cozzi, L. and Hackmann, B. (2022). Realization of Paris Agreement pledges may limit warming just below 2°C. *Nature* 604 (7905): 304–309.

Natali, S. M., Holdren, J. P., Rogers, B. M., Treharne, R., Duffy, P. B., Pomerance, R. and MacDonald, E. (2021). Permafrost carbon feedbacks threaten global climate goals. *Proceedings of the National Academy of Sciences* 118 (21): e2100163118.

Nicholls, R. J., Lincke, D., Hinkel, J., Brown, S., Vafeidis, A. T., Meyssignac, B., Hanson, S. E., Merkens, J.-L. and Fang, J. (2021). A global analysis of subsidence, relative sea-level change and coastal flood exposure. *Nature Climate Change* 11 (4): 338–342.

O'Neill, B. C., Tebaldi, C., van Vuuren, D. P., Eyring, V., Friedlingstein, P., Hurtt, G., Knutti, R., Kriegler, E., Lamarque, J. F., Lowe, J., Meehl, G. A., Moss, R., Riahi, K. and Sanderson, B. M. (2016). The Scenario Model Intercomparison Project (ScenarioMIP) for CMIP6. *Geoscientific Model Development* 9 (9): 3461–3482.

Past Interglacials Working Group of PAGES (2016). Interglacials of the last 800,000 years. *Reviews of Geophysics* 54 (1): 2015RG000482.

Persson, L., Carney Almroth, B. M., Collins, C. D., Cornell, S., de Wit, C. A., Diamond, M. L., Fantke, P., Hassellöv, M., MacLeod, M., Ryberg, M. W., Søgaard Jørgensen, P., Villarrubia-Gómez, P., Wang, Z. and Hauschild, M. Z. (2022). Outside the safe operating space of the planetary boundary for novel entities. *Environmental Science & Technology* 56 (3): 1510–1521.

Rasmussen, S. O., Bigler, M., Blockley, S. P., Blunier, T., Buchardt, S. L., Clausen, H. B., Cvijanovic, I., Dahl-Jensen, D., Johnsen, S. J., Fischer, H., Gkinis, V., Guillevic, M., Hoek, W. Z., Lowe, J. J., Pedro, J. B., Popp, T., Seierstad, I. K., Steffensen, J. P., Svensson, A. M., Vallelonga, P., Vinther, B. M., Walker, M. J. C., Wheatley, J. J. and Winstrup, M. (2014). A stratigraphic framework for abrupt climatic changes during the Last Glacial period based on three synchronized Greenland ice-core records: refining and extending the INTIMATE event stratigraphy. *Quaternary Science Reviews* 106: 14–28.

Readfearn, G. (2022). Great Barrier Reef authority confirms unprecedented sixth mass coral bleaching event. *The Guardian*, 25 March 2022. Accesed at: https://www.theguardian.com/environment/2022/mar/25/we-need-action-immediately-great-barrier-reef-authority-confirms-sixth-mass-coral-bleaching-event.

Riahi, K., van Vuuren, D. P., Kriegler, E., Edmonds, J., O'Neill, B. C., Fujimori, S., Bauer, N., Calvin, K., Dellink, R., Fricko, O., Lutz, W., Popp, A., Cuaresma, J. C., Kc, S., Leimbach, M., Jiang, L., Kram, T., Rao, S., Emmerling, J., Ebi, K., Hasegawa, T., Havlik, P., Humpenöder, F., Da Silva, L. A., Smith, S., Stehfest, E., Bosetti, V., Eom, J., Gernaat, D., Masui, T., Rogelj, J., Strefler, J., Drouet, L., Krey, V., Luderer, G., Harmsen, M., Takahashi, K., Baumstark, L., Doelman, J. C., Kainuma, M., Klimont, Z., Marangoni, G., Lotze-Campen, H., Obersteiner, M., Tabeau, A. and Tavoni, M. (2017). The Shared Socioeconomic Pathways and their energy, land use, and greenhouse gas emissions implications: an overview. *Global Environmental Change* 42: 153–168.

Ritchie, P. D. L., Smith, G. S., Davis, K. J., Fezzi, C., Halleck-Vega, S., Harper, A. B., Boulton, C. A., Binner, A. R., Day, B. H., Gallego-Sala, A. V., Mecking, J. V., Sitch, S. A., Lenton, T. M. and Bateman, I. J. (2020). Shifts in national land use and food production in Great Britain after a climate tipping point. *Nature Food* 1 (1): 76–83.

Rockström, J., Steffen, W., Noone, K., Persson, Å., Chapin, F. S., Lambin, E. F., Lenton, T. M., Scheffer, M., Folke, C., Schellnhuber, H. J., Nykvist, B., de Wit, C. A., Hughes, T., van der Leeuw, S., Rodhe, H., Sörlin, S., Snyder, P. K., Costanza, R., Svedin, U., Falkenmark, M., Karlberg, L., Corell, R. W., Fabry, V. J., Hansen, J., Walker, B., Liverman, D., Richardson, K., Crutzen, P. and Foley, J. A. (2009). A safe operating space for humanity. *Nature* 461: 472–475.

Schuur, E. A. G., McGuire, A. D., Schädel, C., Grosse, G., Harden, J. W., Hayes, D. J., Hugelius, G., Koven, C. D., Kuhry, P., Lawrence, D. M., Natali, S. M., Olefeldt, D., Romanovsky, V. E., Schaefer, K., Turetsky, M. R., Treat, C. C. and Vonk, J. E. (2015). Climate change and the permafrost carbon feedback. *Nature* 520 (7546): 171–179.

Schweiger, A. J., Steele, M., Zhang, J., Moore, G. W. K. and Laidre, K. L. (2021). Accelerated sea ice loss in the Wandel Sea points to a change in the Arctic's Last Ice Area. *Communications Earth & Environment* 2 (1): 122.

Steffen, W., Broadgate, W., Deutsch, L., Gaffney, O. and Ludwig, C. (2015). The trajectory of the Anthropocene: the Great Acceleration. *The Anthropocene Review* 2 (1): 81–98.

Steffen, W., Richardson, K., Rockström, J., Cornell, S. E., Fetzer, I., Bennett, E. M., Biggs, R., Carpenter, S. R., de Vries, W., de Wit, C. A., Folke, C., Gerten, D., Heinke, J., Mace, G. M., Persson, L. M., Ramanathan, V., Reyers, B. and Sörlin, S. (2015). Planetary boundaries: guiding human development on a changing planet. *Science* 347 (6223): 1259855.

Steffensen, J. P., Andersen, K. K., Bigler, M., Clausen, H. B., Dahl-Jensen, D., Fischer, H., Goto-Azuma, K., Hansson, M., Johnsen, S. J., Jouzel, J., Masson-Delmotte, V., Popp, T., Rasmussen, S. O., Röthlisberger, R., Ruth, U., Stauffer, B., Siggaard-Andersen, M.-L., Sveinbjörnsdóttir, Á. E., Svensson, A. and White, J. W. C. (2008). High-resolution Greenland ice core data show abrupt climate change happens in few years. *Science* 321 (5889): 680–684.

Suarez-Gutierrez, L., Müller, W. A., Li, C. and Marotzke, J. (2020). Hotspots of extreme heat under global warming. *Climate Dynamics* 55 (3): 429–447.

Trewin, B., Cazenave, A., Howell, S., Huss, M., Isensee, K., Palmer, M. D., Tarasova, O. and Vermeulen, A. (2021). Headline indicators for global climate monitoring. *Bulletin of the American Meteorological Society* 102 (1): E20–E37.

United Nations Environment Programme (2021). The Emissions Gap Report 2021. UNEP, Nairobi, Kenya. Accessed at: https://www.unep.org/resources/emissions-gap-report-2021.

United Nations Secretary-General (2022). Secretary-General's video message to the Press Conference Launch of IPCC Report. Accessed at: https://www.un.org/sg/en/node/262102.

van Oldenborgh, G. J., van der Wiel, K., Kew, S., Philip, S., Otto, F., Vautard, R., King, A., Lott, F., Arrighi, J., Singh, R. and van Aalst, M. (2021). Pathways and pitfalls in extreme event attribution. *Climatic Change* 166 (1): 13.

World Meteorological Organization (2021). *State of Climate in 2021: WMO Provisional Report.* Accessed at: https://library.wmo.int/index.php?lvl=notice_display&id=21982#.Yd-bCS0RrUI.

PART 2

Albert, S., Leon, J. X., Grinham, A. R., Church, J. A., Gibbes, B. R. and Woodroffe, C. D. (2016). Interactions between sea-level rise and wave exposure on reef island dynamics in the Solomon Islands. *Environmental Research Letters* 11 (5): 054011.

Atkinson, J. (2021). *Facing It: A Podcast About Love, Loss, and the Natural World.* Accessed at: https://www.drjenniferatkinson.com/facing-it.

Attenborough, D. (2020). *A Life on Our Planet: My Witness Statement and Vision for the Future.* Penguin Random House, London, UK.

Boer, M. M., Resco de Dios, V. and Bradstock, R. A. (2020). Unprecedented burn area of Australian mega forest fires. *Nature Climate Change* 10: 171–172.

Carbon Brief (2021). Will global warming 'stop' as soon as net-zero emissions are reached? Accessed at: https://www.carbonbrief.org/explainer-will-global-warming-stop-as-soon-as-net-zero-emissions-are-reached.

Connell, J. (2016). Last days in the Carteret Islands? Climate change, livelihoods and migration on coral atolls. *Asia Pacific Viewpoint* 57 (1): 3–15.

Crowther, T. W., Glick, H. B., Covey, K. R., Bettigole, C., Maynard, D. S., Thomas, S. M., Smith, J. R., Hintler, G., Duguid, M. C., Amatulli, G., Tuanmu, M. N., Jetz, W., Salas, C., Stam, C., Piotto, D., Tavani, R., Green, S., Bruce, G., Williams, S. J., Wiser, S. K., Huber, M. O., Hengeveld, G. M., Nabuurs, G. J., Tikhonova, E., Borchardt, P., Li, C. F., Powrie, L. W., Fischer, M., Hemp, A., Homeier, J., Cho, P., Vibrans, A. C., Umunay, P. M., Piao, S. L., Rowe, C. W., Ashton, M. S., Crane, P. R. and Bradford, M. A. (2015). Mapping tree density at a global scale. *Nature* 525 (7568): 201–205.

Davidson, N. C. (2014). How much wetland has the world lost? Long-term and recent trends in global wetland area. *Marine and Freshwater Research* 65 (10): 934–941.

Davis, K. F., Bhattachan, A., D'Odorico, P. and Suweis, S. (2018). A universal model for predicting human migration under climate change: examining future sea level rise in Bangladesh. *Environmental Research Letters* 13 (6): 064030.

Dosio, A., Mentaschi, L., Fischer, E. M. and Wyser, K. (2018). Extreme heat waves under 1.5°C and 2°C global warming. *Environmental Research Letters* 13 (5): 054006.

Doyle, G. (2016). *Love Warrior: A Memoir.* Flatiron Books, New York, USA.

Economic and Social Commission for Asia and the Pacific (2014). Climate Change and Migration Issues in the Pacific. Accessed at: https://www.ilo.org/dyn/migpractice/docs/261/Pacific.pdf.

Eddy, T. D., Lam, V. W. Y., Reygondeau, G., Cisneros-Montemayor, A. M., Greer, K., Palomares, M. L. D., Bruno, J. F., Ota, Y. and Cheung, W. W. L. (2021). Global decline in capacity of coral reefs to provide ecosystem services. *One Earth* 4 (9): 1278–1285.

Energy and Climate Intelligence Unit (2022). Net Zero Scorecard 2022. Accessed at: https://eciu.net/netzerotracker.

Farbotko, C., Dun, O., Thornton, F., McNamara, K. E. and McMichael, C. (2020). Relocation planning must address voluntary immobility. *Nature Climate Change* 10 (8): 702–704.

Filkov, A. I., Ngo, T., Matthews, S., Telfer, S. and Penman, T. D. (2020). Impact of Australia's catastrophic 2019/20 bushfire season on communities and environment: retrospective analysis and current trends. *Journal of Safety Science and Resilience* 1 (1): 44–56.

Food and Agriculture Organization and United Nations Environment Programme (2020). The State of the World's Forests 2020. Accessed at: https://www.fao.org/3/ca8642en/ca8642en.pdf.

Francis, P. (2015). Encyclical letter *Laudato Si: On Care for Our Common Home.* Accessed at: https://www.vatican.va/content/francesco/en/encyclicals/documents/papa-francesco_20150524_enciclica-laudato-si.html.

Friedlingstein, P., Jones, M. W., O'Sullivan, M., Andrew, R. M., Bakker, D. C. E., Hauck, J., Le Quéré, C., Peters, G. P., Peters, W., Pongratz, J., Sitch, S., Canadell, J. G., Ciais, P., Jackson, R. B., Alin, S. R., Anthoni, P., Bates, N. R., Becker, M., Bellouin, N., Bopp, L., Chau, T. T. T., Chevallier, F., Chini, L. P., Cronin, M., Currie, K. I., Decharme, B., Djeutchouang, L., Dou, X., Evans, W., Feely, R. A., Feng, L., Gasser, T., Gilfillan, D., Gkritzalis, T., Grassi, G., Gregor, L., Gruber, N., Gürses, Ö., Harris, I., Houghton, R. A., Hurtt, G. C., Iida, Y., Ilyina, T., Luijkx, I. T., Jain, A. K., Jones, S. D., Kato, E., Kennedy, D., Klein Goldewijk, K., Knauer, J., Korsbakken, J. I., Körtzinger, A., Landschützer, P., Lauvset, S. K., Lefèvre, N., Lienert, S., Liu, J., Marland, G., McGuire, P. C., Melton, J. R., Munro, D. R., Nabel, J. E. M. S., Nakaoka, S. I., Niwa, Y., Ono, T., Pierrot, D., Poulter, B., Rehder, G., Resplandy, L., Robertson, E., Rödenbeck, C., Rosan, T. M., Schwinger, J., Schwingshackl, C., Séférian, R., Sutton, A. J., Sweeney, C., Tanhua, T., Tans, P. P., Tian, H., Tilbrook, B., Tubiello, F., van der Werf, G., Vuichard, N., Wada, C., Wanninkhof, R., Watson, A., Willis, D., Wiltshire, A. J., Yuan, W., Yue, C., Yue, X., Zaehle, S. and Zeng, J. (2022). Global Carbon Budget 2021. *Earth System Science Data* 14: 1917–2005.

García-Herrera, R., Díaz, J., Trigo, R. M., Luterbacher, J. and Fischer, E. M. (2010). A review of the European summer heat wave of 2003. *Critical Reviews in Environmental Science and Technology* 40 (4): 267–306.

Gard, R. and Veitayaki, J. (2017). In the wake of Winston: climate change, mobility and resiliency in Fiji. *International Journal of Safety and Security Engineering* 7 (2): 157–168.

Global Carbon Capture and Storage Institute (2021). Global Status of CCS Report 2021. Accessed at: https://www.globalccsinstitute.com/resources/global-status-report/.

Guo, Y., Wu, Y., Wen, B., Huang, W., Ju, K., Gao, Y. and Li, S. (2020). Floods in China, COVID-19, and climate change. *The Lancet Planetary Health* 4 (10): e443–e444.

Hauʻofa, E. (1998). The ocean in us. *The Contemporary Pacific* 10 (2): 391–410.

Hickman, C., Marks, E., Pihkala, P., Clayton, S., Lewandowski, E., Mayall, E., Wray, B., Mellor, C. and van Susteren, L. (2021). Young people's voices on climate anxiety, government betrayal and moral injury: a global phenomenon. *The Lancet*: http://dx.doi.org/10.2139/ssrn.3918955.

Hooijer, A. and Vernimmen, R. (2021). Global LiDAR land elevation data reveal greatest sea-level rise vulnerability in the tropics. *Nature Communications* 12 (1): 3592.

Hu, X., Huang, B., Verones, F., Cavalett, O. and Cherubini, F. (2021). Overview of recent land-cover changes in biodiversity hotspots. *Frontiers in Ecology and the Environment* 19 (2): 91–97.

Hughes, T. P., Kerry, J. T., Connolly, S. R., Álvarez-Romero, J. G., Eakin, C. M., Heron, S. F., Gonzalez, M. A. and Moneghetti, J. (2021). Emergent properties in the responses of tropical corals to recurrent climate extremes. *Current Biology* 31 (23): https://doi.org/10.1016/j.cub.2021.10.046.

Internal Displacement Monitoring Centre (2021). *Global Report on Internal Displacement: Internal Displacement in a Changing Climate*. Accessed at: https://www.internal-displacement.org/global-report/grid2021/.

International Cryosphere Climate Initiative (2021). *State of the Cryosphere 2021: A Needed Decade of Urgent Action*. International Cryosphere Climate Initiative. Accessed at: http://iccinet.org/statecryo21/.

IPBES (2019). *Global Assessment Report on Biodiversity and Ecosystem Services of the Intergovernmental Science-Policy Platform on Biodiversity and Ecosystem Services*. Brondizio, E. S., Settele, J., Díaz, S. and Ngo, H.T. (eds). IPBES Secretariat, Bonn, Germany. Accessed at: https://doi.org/10.5281/zenodo.3831673.

IPBES–IPCC (2021). *Biodiversity and Climate Change: Scientific Outcome of the IPBES–IPCC Co-sponsored Workshop on Biodiversity and Climate Change*. Accessed at: https://ipbes.net/sites/default/files/2021-06/2021_IPCC-IPBES_scientific_outcome_20210612.pdf.

IPCC (2018). *Global Warming of 1.5°C: An IPCC Special Report on the Impacts of Global Warming of 1.5°C Above Pre-Industrial Levels and Related Global Greenhouse Gas Emission Pathways, in the Context of Strengthening the Global Response to the Threat of Climate Change, Sustainable Development, and Efforts to Eradicate Poverty*. Masson-Delmotte, V., Zhai, P., Pörtner, H.-O., Roberts, D., Skea, J., Shukla, P. R., Pirani, A., Moufouma-Okia, W., Péan, C., Pidcock, R., Connors, S., Matthews, J. B. R., Chen, Y., Zhou, X., Gomis, M. I., Lonnoy, E., Maycock, T., Tignor, M. and Waterfield, T. (eds). Cambridge University Press, Cambridge, UK. Accessed at: https://www.ipcc.ch/sr15/.

IPCC (2019a). *IPCC Special Report on the Ocean and Cryosphere in a Changing Climate.* Pörtner, H. O., Roberts, D. C., Masson-Delmotte, V., Zhai, P., Tignor, M., Poloczanska, E., Mintenbeck, K., Nicolai, M., Okem, A. and Petzold, J. (eds). Cambridge University Press, Cambridge, UK. Accessed at: https://www.ipcc.ch/srocc/.

IPCC (2019b). Jia, G., Shevliakova, E., Artaxo, P., De Noblet-Ducoudré, N., Houghton, R., House, J., Kitajima, K., Lennard, C., Popp, A., Sirin, A., Sukumar, R. and Verchot, L. (2019b). Land-climate interactions. In *Climate Change and Land: An IPCC Special Report on Climate Change, Desertification, Land Degradation, Sustainable Land Management, Food Security, and Greenhouse Gas Fluxes in Terrestrial Ecosystems.* Shukla, J. S. P.R., Calvo Buendia, E., Masson-Delmotte, V., Pörtner, H.-O., Roberts, D. C., Zhai, P., Slade, R., Connors, S., van Diemen, R., Ferrat, M., Haughey, E., Luz, S., Neogi, S., Pathak, Petzold, J., Portugal Pereira, J., Vyas, P., Huntley, E., Kissick, K., Belkacemi, M., Malley, J. (eds). Cambridge University Press, Cambridge, UK.

IPCC (2021a). IPCC Working Group I Interactive Atlas. Accessed at: https://interactive-atlas.ipcc.ch.

IPCC (2021b). Summary for Policymakers. In *Climate Change 2021: The Physical Science Basis. Contribution of Working Group I to the Sixth Assessment Report of the Intergovernmental Panel on Climate Change.* Masson-Delmotte, V., Zhai, P., Pirani, A., Connors, S. L., Péan, C., Berger, S., Caud, N., Chen, Y., Goldfarb, L., Gomis, M. I., Huang, M., Leitzell, K., Lonnoy, E., Matthews, J. B. R., Maycock, T. K., Waterfield, T., Yelekçi, O., Yu, R. and Zhou, B. (eds). Accessed at: https://www.ipcc.ch/report/ar6/wg1/.

IPCC (2022a). *Climate Change 2022: Impacts, Adaptation and Vulnerability. Contribution of Working Group II to the Sixth Assessment Report of the Intergovernmental Panel on Climate Change.* Accessed at: https://www.ipcc.ch/report/ar6/wg2/.

IPCC (2022b). Technical Summary. In *Climate Change 2022: Mitigation of Climate Change, the Working Group III Contribution to the Sixth Assessment Report of the Intergovernmental Panel on Climate Change.* Masson-Delmotte, V., Zhai, P., Pirani, A., Connors, S. L., Péan, C., Berger, S., Caud, N., Chen, Y., Goldfarb, L., Gomis, M. I., Huang, M., Leitzell, K., Lonnoy, E., Matthews, J. B. R., Maycock, T. K., Waterfield, T., Yelekçi, O., Yu, R. and Zhou, B. (eds). Accessed at: https://www.ipcc.ch/report/ar6/wg3/.

IPCC, Bednar-Friedl, B., Biesbroek, R., Schmidt, D. N., Alexander, P., Borsheim, K. Y., Carnicer, J., Georgopoulou, E., Haasnoot, M., Cozannet, G., Lionello, P., Lipka, O., Mollmann, C., Muccione, V., Mustonen, T., Piepenburg, D. and Whitmarsh, L. (2022a). Chapter 13: Europe. In *Climate Change 2022: Impacts, Adaptation and Vulnerability. Contribution of Working Group II to the Sixth*

Assessment Report of the Intergovernmental Panel on Climate Change. Accessed at: https://www.ipcc.ch/report/ar6/wg2/.

IPCC, Caretta, M. A., Mukherji, A., Arfanuzzaman, M., Betts, R. A., Gelfan, A., Hirabayashi, Y., Lissner, T. K., Lopez Gunn, E., Liu, J., Morgan, R., Mwanga, S. and Supratid, S. (2022b). Chapter 4: Water. In *Climate Change 2022: Impacts, Adaptation and Vulnerability. Contribution of Working Group II to the Sixth Assessment Report of the Intergovernmental Panel on Climate Change.* Accessed at: https://www.ipcc.ch/report/ar6/wg2/.

IPCC, Costello, M. J., Vale, M. M., Kiessling, W., Maharaj, S., Price, J. and Talukdar, G. H. (2022c). Cross-chapter Paper 1: Biodiversity Hotspots. In *Climate Change 2022: Impacts, Adaptation and Vulnerability. Contribution of Working Group II to the Sixth Assessment Report of the Intergovernmental Panel on Climate Change.* Accessed at: https://www.ipcc.ch/report/ar6/wg2/.

IPCC, Dhakal, S., Minx, J. C., Toth, F. L., Abdel-Aziz, A., Figueroa–Meza, M. J., Hubacek, K., Jonckheere, I., Kim, Y. G., Nemet, G. F., Pachauri, S., Tan, X. and Wiedmann, T. (2022d). Chapter 2: Emissions Trends and Drivers. In *Climate Change 2022: Mitigation of Climate Change Contribution of Working Group III to the Sixth Assessment Report of the Intergovernmental Panel on Climate Change.* Accessed at: https://www.ipcc.ch/report/ar6/wg3/.

IPCC, Dodman, D., Hayward, B., Pelling, M., Castan Broto, V., Winston Chow, W., Chu, E., Dawson, R., Khirfan, L., McPhearson, T., Prakash, A., Zheng, Y. and Ziervogel, G. (2022e). Chapter 6: Cities, Settlements and Key Infrastructure. In *Climate Change 2022: Impacts, Adaptation and Vulnerability. Contribution of Working Group II to the Sixth Assessment Report of the Intergovernmental Panel on Climate Change.* Accessed at: https://www.ipcc.ch/report/ar6/wg2/.

IPCC, Glavovic, B., Dawson, R., Chow, W., Garschagen, M., Haasnoot, M., Singh, C. and Thomas, A. (2022f). Cross-Chapter Paper 2: Cities and Settlements by the Sea. In *Climate Change 2022: Impacts, Adaptation and Vulnerability. Contribution of Working Group II to the Sixth Assessment Report of the Intergovernmental Panel on Climate Change.* Accessed at: https://www.ipcc.ch/report/ar6/wg2/.

IPCC, Mycoo, M., Wairiu, M., Campbell, D., Duvat, V., Golbuu, Y., Maharaj, S., Nalau, J., Nunn, P., Pinnegar, J. and Warrick, O. (2022g). Chapter 15: Small Islands. In *Climate Change 2022: Impacts, Adaptation and Vulnerability. Contribution of Working Group II to the Sixth Assessment Report of the Intergovernmental Panel on Climate Change.* Accessed at: https://www.ipcc.ch/report/ar6/wg2/.

IPCC, Ometto, J. P., Kalaba, F. K., Anshari, G. Z., Chacon, N., Farrell, A., Abdul Halim, S., Neufeldt, H. and Raman Sukumar, R. (2022h). Cross-Chapter Paper 7: Tropical Forests. In *Climate Change 2022: Impacts, Adaptation and Vulnerability. Contribution of Working Group II to the Sixth Assessment Report of the Intergovernmental Panel on Climate Change.* Accessed at: https://www.ipcc.ch/report/ar6/wg2/.

IPCC, Parmesan, C., Morecroft, M. D., Yongyut Trisurat, Y., Adrian, R., Zakaria Anshari, G., Arneth, A., Gao, Q., Gonzalez, P., Harris, R., Price, J., Stevens, N. and Hirak Talukdar, G. (2022i). Chapter 2: Terrestrial and Freshwater Ecosystems and their Services. In *Climate Change 2022: Impacts, Adaptation and Vulnerability. Contribution of Working Group II to the Sixth Assessment Report of the Intergovernmental Panel on Climate Change*. Accessed at: https://www.ipcc.ch/report/ar6/wg2/.

IPCC, Seneviratne, S. I., Zhang, X., Adnan, M., Badi, W., Dereczynski, C., Di Luca, A., Ghosh, S., Iskandar, I., Kossin, J., Lewis, S., Otto, F., Pinto, I., Satoh, M., Vicente-Serrano, S. M., Wehner, M. and Zhou, B. (2021). Chapter 11: Weather and Climate Extreme Events in a Changing Climate. In *Climate Change 2021: The Physical Science Basis. Contribution of Working Group I to the Sixth Assessment Report of the Intergovernmental Panel on Climate Change*. Masson-Delmotte, V., Zhai, P., Pirani, A., Connors, S. L., Péan, C., Berger, S., Caud, N., Chen, Y., Goldfarb, L., Gomis, M. I., Huang, M., Leitzell, K., Lonnoy, E., Matthews, J. B. R., Maycock, T. K., Waterfield, T., Yelekçi, O., Yu, R. and Zhou, B. (eds). Accessed at: https://www.ipcc.ch/report/ar6/wg1/.

IPCC, Schipper, E. L. F., Aromar Revi, A., Preston, B. L., Carr, E. R., Eriksen, S. H., Fernandez-Carril, L. R., Glavovic, B., Hilmi, N. J. M., Ley, D., Mukerji, R., Muylaert de Araujo, M. S., Perez, R., Rose, S. K. and Singh, P. K. (2022j). Chapter 18: Climate Resilient Development Pathways. In *Climate Change 2022: Impacts, Adaptation and Vulnerability. Contribution of Working Group II to the Sixth Assessment Report of the Intergovernmental Panel on Climate Change*. Accessed at: https://www.ipcc.ch/report/ar6/wg2/.

IPCC, Shaw, R., Luo, Y., Cheong, T. S., Abdul Halim, S., Chaturvedi, S., Hashizume, M., Insarov, G. E., Ishikawa, Y., Jafari, M., Kitoh, A., Pulhin, J., Singh, C., Vasant, K. and Zhibin Zhang, Z. (2022k). Chapter 10: Asia. In *Climate Change 2022: Impacts, Adaptation and Vulnerability. Contribution of Working Group II to the Sixth Assessment Report of the Intergovernmental Panel on Climate Change*. Accessed at: https://www.ipcc.ch/report/ar6/wg2/.

IPCC, Canadell, J. G., Monteiro, P. M. S., Costa, M. H., Cotrim da Cunha, L., Cox, P. M., Eliseev, A. V., Henson, S., Ishii, M., Jaccard, S., Koven, C., Lohila, A., Patra, P. K., Piao, S., Rogelj, J., Syampungani, S., Zaehle, S., and Zickfeld, K. (2021). Chapter 5: Global Carbon and Other Biogeochemical Cycles and Feedbacks Supplementary Material. In *Climate Change 2021: The Physical Science Basis. Contribution of Working Group I to the Sixth Assessment Report of the Intergovernmental Panel on Climate Change*. Masson-Delmotte, V., Zhai, P., Pirani, A., Connors, S. L., Péan, C., Berger, S., Caud, N., Chen, Y., Goldfarb, L., Gomis, M. I., Huang, M., Leitzell, K., Lonnoy, E., Matthews, J. B. R., Maycock, T. K., Waterfield, T., Yelekçi, O., Yu, R. and Zhou, B. (eds). Accessed at: https://www.ipcc.ch/report/ar6/wg1/.

Klein Salamon, M. and Gage, M. (2020). *Facing the Climate Emergency: How to Transform Yourself with Climate Truth.* New Society Publishers, Gabriola Island, Canada.

Maragos, J. E. and Williams, G. J. (2011). Pacific Coral Reefs: An Introduction. *Encyclopedia of Modern Coral Reefs: Structure, Form and Process.* Hopley, D (ed.). Springer Netherlands, Dordrecht: 753-776.

Matthews, T. K. R., Wilby, R. L. and Murphy, C. (2017). Communicating the deadly consequences of global warming for human heat stress. *Proceedings of the National Academy of Sciences* 114 (15): 3861.

Meinshausen, M., Lewis, J., McGlade, C., Gütschow, J., Nicholls, Z., Burdon, R., Cozzi, L. and Hackmann, B. (2022). Realization of Paris Agreement pledges may limit warming just below 2°C. *Nature* 604 (7905): 304-309.

Merkens, J.-L., Reimann, L., Hinkel, J. and Vafeidis, A. T. (2016). Gridded population projections for the coastal zone under the Shared Socioeconomic Pathways. *Global and Planetary Change* 145: 57-66.

Miner, K. R., Turetsky, M. R., Malina, E., Bartsch, A., Tamminen, J., McGuire, A. D., Fix, A., Sweeney, C., Elder, C. D. and Miller, C. E. (2022). Permafrost carbon emissions in a changing Arctic. *Nature Reviews Earth & Environment* 3 (1): 55-67.

Mitchell, D., Heaviside, C., Vardoulakis, S., Huntingford, C., Masato, G., Guillod, B. P., Frumhoff, P., Bowery, A., Wallom, D. and Allen, M. (2016). Attributing human mortality during extreme heat waves to anthropogenic climate change. *Environmental Research Letters* 11 (7): 074006.

Monbiot, G. (2021). *This Can't Be Happening.* Penguin Random House, London, UK.

Mora, C., Dousset, B., Caldwell, I. R., Powell, F. E., Geronimo, R. C., Bielecki, Coral R., Counsell, C. W. W., Dietrich, B. S., Johnston, E. T., Louis, L. V., Lucas, M. P., McKenzie, M. M., Shea, A. G., Tseng, H., Giambelluca, T. W., Leon, L. R., Hawkins, E. and Trauernicht, C. (2017). Global risk of deadly heat. *Nature Climate Change* 7: 501.

Myers, R. A. and Worm, B. (2003). Rapid worldwide depletion of predatory fish communities. *Nature* 423 (6937): 280-283.

Net Zero Tracker (2022). Net Zero Tracker. Accessed at: https://zerotracker.net.

Nolan, R. H., Bowman, D. M. J. S., Clarke, H., Haynes, K., Ooi, M. K. J., Price, O. F., Williamson, G. J., Whittaker, J., Bedward, M., Boer, M. M., Cavanagh, V. I., Collins, L., Gibson, R. K., Griebel, A., Jenkins, M. E., Keith, D. A., McIlwee, A. P., Penman, T. D., Samson, S. A., Tozer, M. G. and Bradstock, R. A. (2021). What do the Australian Black Summer fires signify for the Global Fire Crisis? *Fire* 4 (4): https://doi.org/10.3390/fire4040097.

Nunn, P. D., Kohler, A. and Kumar, R. (2017). Identifying and assessing evidence for recent shoreline change attributable to uncommonly rapid sea-level rise in Pohnpei, Federated States of Micronesia, Northwest Pacific Ocean. *Journal of Coastal Conservation* 21 (6): 719-730.

Oxfam (2020). Confronting Carbon Inequality: Putting Climate Justice at the Heart of the COVID-19 Recovery. Oxfam International. Accessed at: https://policy-practice.oxfam.org/resources/confronting-carbon-inequality-putting-climate-justice-at-the-heart-of-the-covid-621052/.

Piggott-McKellar, A. E., McNamara, K. E., Nunn, P. D. and Sekinini, S. T. (2019). Moving people in a changing climate: lessons from two case studies in Fiji. Social Sciences 8 (5): https://doi.org/10.3390/socsci8050133.

Plumptre, A. J., Baisero, D., Belote, R. T., Vázquez-Domínguez, E., Faurby, S., Jędrzejewski, W., Kiara, H., Kühl, H., Benítez-López, A., Luna-Aranguré, C., Voigt, M., Wich, S., Wint, W., Gallego-Zamorano, J. and Boyd, C. (2021). Where might we find ecologically intact communities? Frontiers in Forests and Global Change 4 (26): DOI 10.3389/ffgc.2021.626635.

Sharmila, S. and Walsh, K. J. E. (2018). Recent poleward shift of tropical cyclone formation linked to Hadley cell expansion. Nature Climate Change 8: 730-736

Sherwood, S. C. and Huber, M. (2010). An adaptability limit to climate change due to heat stress. Proceedings of the National Academy of Sciences 107 (21): 9552.

Stott, P. A., Stone, D. A. and Allen, M. R. (2004). Human contribution to the European heatwave of 2003. Nature 432 (7017): 610-614.

Thiery, W., Lange, S., Rogelj, J., Schleussner, C.-F., Gudmundsson, L., Seneviratne Sonia, I., Andrijevic, M., Frieler, K., Emanuel, K., Geiger, T., Bresch David, N., Zhao, F., Willner Sven, N., Büchner, M., Volkholz, J., Bauer, N., Chang, J., Ciais, P., Dury, M., François, L., Grillakis, M., Gosling Simon, N., Hanasaki, N., Hickler, T., Huber, V., Ito, A., Jägermeyr, J., Khabarov, N., Koutroulis, A., Liu, W., Lutz, W., Mengel, M., Müller, C., Ostberg, S., Reyer Christopher, P. O., Stacke, T. and Wada, Y. (2021). Intergenerational inequities in exposure to climate extremes. Science 374 (6564): 158-160.

Tusting, L. S., Bisanzio, D., Alabaster, G., Cameron, E., Cibulskis, R., Davies, M., Flaxman, S., Gibson, H. S., Knudsen, J., Mbogo, C., Okumu, F. O., von Seidlein, L., Weiss, D. J., Lindsay, S. W., Gething, P. W. and Bhatt, S. (2019). Mapping changes in housing in sub-Saharan Africa from 2000 to 2015. Nature 568 (7752): 391-394.

UN Habitat (2016). Slum Almanac 2015-2016. UN Habitat, Nairobi, Kenya. Accessed at: https://unhabitat.org/sites/default/files/documents/2019-05/slum_almanac_2015-2016_psup.pdf.

UN Habitat (2020). World Cities Report 2020: The Value of Sustainable Urbanization. United Nations Human Settlements Programme, Nairobi, Kenya. Accessed at: https://unhabitat.org/wcr/.

United Nations Environment Programme (2021). Making Peace with Nature: A Scientific Blueprint to Tackle the Climate, Biodiversity and Pollution Emergencies. United Nations Environment Programme, Nairobi, Kenya. Accessed at: https://www.unep.org/resources/making-peace-nature.

United Nations Harmony with Nature Programme (2021). Rights of Nature Law and Policy. Accessed at: http://www.harmonywithnatureun.org/rightsOfNature/.

van Oldenborgh, G. J., Philip, S., Kew, S., van Weele, M., Uhe, P., Otto, F., Singh, R., Pai, I., Cullen, H. and AchutaRao, K. (2018). Extreme heat in India and anthropogenic climate change. *Natural Hazards and Earth System Sciences* 18 (1): 365–381.

Victorian Parliament (2017). *Yarra River Protection (Wilip-gin Birrarung murron) Act 2017.* Accessed at: http://classic.austlii.edu.au/au/legis/vic/consol_act/yrpbma2017554/.

Wells, L. (2021). *Believers: Making a Life at the End of the World.* Farrar, Straus and Giroux, New York, USA.

Welsby, D., Price, J., Pye, S. and Ekins, P. (2021). Unextractable fossil fuels in a 1.5°C world. *Nature* 597 (7875): 230–234.

Wiens, J. J. (2016). Climate-related local extinctions are already widespread among plant and animal species. *PLOS Biology* 14 (12): e2001104.

Wiseman, J. (2021). *Hope and Courage in the Climate Crisis: Wisdom and Action in the Long Emergency.* Palgrave Macmillan, Cham, Switzerland.

World Bank (2021). *Groundswell Part 2: Acting on Internal Climate Migration.* World Bank, Washington, DC, USA. Accessed at: https://climate-diplomacy.org/magazine/conflict/groundswell-part-ii-acting-internal-climate-migration.

World Inequality Lab (2022). *World Inequality Report 2022.* Accessed at: https://wir2022.wid.world.

PART 3

Abrahams, G., Johnson, B. and Gellatly, K. (2016). *ART+CLIMATE= CHANGE.* Melbourne University Press, Melbourne, Australia.

Allam, L. (2020). For First Nations people the bushfires bring a particular grief, burning what makes us who we are. *The Guardian,* 6 January 2020. Accessed at: https://www.theguardian.com/commentisfree/2020/jan/06/for-first-nations-people-the-bushfires-bring-a-particular-grief-burning-what-makes-us-who-we-are.

Andersen, K. G., Rambaut, A., Lipkin, W. I., Holmes, E. C. and Garry, R. F. (2020). The proximal origin of SARS-CoV-2. *Nature Medicine* 26 (4): 450–452.

ARTCOP21 (2015). ARTCOP21: A Global Festival of Cultural Activity on Climate Change. Accessed at: http://www.artcop21.com.

Artists & Climate Change (2021). Artists & Climate Change: Building Earth Connections. Accessed at: https://artistsandclimatechange.com/about/.

Atkinson, J. (2021). *Facing It: A Podcast About Love, Loss, and the Natural World.* Accessed at: https://www.drjenniferatkinson.com/facing-it.

Attenborough, D. (2020). *A Life on Our Planet: My Witness Statement and Vision for the Future.* Penguin Random House, London, UK.

Australia Institute (2021a). *Fossil Fuel Subsidies in Australia*. Accessed at: https://
australiainstitute.org.au/wp-content/uploads/2021/04/P1021-Fossil-fuel-
subsidies-202 0–21-Web.pdf.

Australia Institute (2021b). *Undermining Climate Action the Australian Way*.
Accessed at: https://australiainstitute.org.au/wp-content/uploads/2021/11/
P1163-Undermining-climate-action-the-Australian-way-WEB.pdf.

Biden, J. (2021). Inaugural Address by President Joseph R. Biden, Jr. Accessed at:
https://www.whitehouse.gov/briefing-room/speeches-remarks/2021/01/20/
inaugural-address-by-president-joseph-r-biden-jr/.

Birch, T. (2017). Climate change, recognition and social place-making. *Sydney
Review of Books*, 3 March 2017. Accessed at: https://sydneyreviewofbooks.com/
essay/climate-change-recognition-and-caring-for-country/.

Carbon Tracker (2021). *The Sky's the Limit: Solar and Wind Energy Potential Is 100
Times as Much as Global Energy Demand*. Accessed at: https://carbontracker.org/
reports/the-skys-the-limit-solar-wind/.

Carrington, D. (2018). Avoiding meat and dairy is 'single biggest way' to reduce your
impact on Earth. *The Guardian*, 1 June 2018. Accessed at: https://www.theguard-
ian.com/environment/2018/may/31/avoiding-meat-and-dairy-is-single-
biggest-way-to-reduce-your-impact-on-earth?CMP=Share_iOSApp_Other.

Cassidy, C. (2021). 'Risky levels': Australia is the drunkest country in the world,
survey finds. *The Guardian*, 3 December 2021. Accessed at: https://www.
theguardian.com/australia-news/2021/dec/03/risky-levels-australia-is-the-
drunkest-country-in-the-world-new-survey-finds.

Centola, D., Becker, J., Brackbill, D. and Baronchelli, A. (2018). Experimental
evidence for tipping points in social convention. *Science* 360 (6393): 1116–1119.

Clean Energy Council (2021). *Clean Energy Australia Report 2021*. Accessed at:
https://www.cleanenergycouncil.org.au/resources/resources-hub/clean-
energy-australia-report.

CLIMARTE (2021). CLIMARTE Emergency: Action. Accessed at:
https://climarte.org.

Climate 200 (2021). Welcome to Climate 200. Accessed at:
https://www.climate200.com.au.

Climate Accountability Institute (2020). Carbon Majors. Accessed at:
https://climateaccountability.org/carbonmajors.html.

Climate Fringe COP26 (2021). Climate Fringe. Accessed at:
https://climatefringe.org.

Climate Mayors (2021). Climate Mayors. Accessed at: https://climatemayors.org.

Convention of Biological Diversity (2021). First draft of the post-2020 Global
Biodiversity Framework. Accessed at: https://www.cbd.int/doc/c/634e/15be/78
d817a6d4ef4520408ed501/wg202 0–03-l-01-en.pdf.

Cook, J. (2016). A brief history of fossil-fuelled climate denial. *The Conversation*, 21 June 2016. Accessed at: https://theconversation.com/a-brief-history-of-fossil-fuelled-climate-denial-61273.

CSIRO (2021). *GenCost 2020–21*. Accessed at: https://www.csiro.au/en/news/News-releases/2021/CSIRO-report-confirms-renewables-still-cheapest-new-build-power-in-Australia.

Daniel, Z. (2021). Zoe Daniel's campaign launch speech, 27 November 2021. Accessed at: https://www.zoedaniel.com.au/2021/11/27/zoe-first-speech/.

Davies, A. (2019a). 'And this is Mosman!': genteel Warringah rocked by election rancour. *The Guardian*, 16 May 2019. Accessed at: https://www.theguardian.com/australia-news/2019/may/16/genteel-warringah-rocked-election-rancour-abbott-steggall.

Davies, A. (2019b). It's the grassroots, stupid: what Zali Steggall's campaign can teach Labor about winning. *The Guardian*, 21 May 2019. Accessed at: https://www.theguardian.com/australia-news/2019/may/21/its-the-grassroots-stupid-what-zali-steggalls-campaign-can-teach-labor-about-winning.

Davies, A. and Cox, L. (2021). Independents' day: why safe Coalition seats are facing grassroots challenges. *The Guardian*, 21 November 2021. Accessed at: https://www.theguardian.com/australia-news/2020/nov/21/independents-day-why-safe-coalition-seats-are-facing-grassroots-challenges.

Di Marco, M., Baker, M. L., Daszak, P., De Barro, P., Eskew, E. A., Godde, C. M., Harwood, T. D., Herrero, M., Hoskins, A. J., Johnson, E., Karesh, W. B., Machalaba, C., Garcia, J. N., Paini, D., Pirzl, R., Smith, M. S., Zambrana-Torrelio, C. and Ferrier, S. (2020). Opinion: Sustainable development must account for pandemic risk. *Proceedings of the National Academy of Sciences* 117 (8): 3888.

EAT-Lancet Commission (2021). The EAT-Lancet Commission on Food, Planet, Health. Accessed at: https://eatforum.org/eat-lancet-commission/.

Energy & Climate Intelligence Unit (2022). Net Zero Scorecard 2022. Accessed at: https://eciu.net/netzerotracker.

Friedlingstein, P., Jones, M. W., O'Sullivan, M., Andrew, R. M., Bakker, D. C. E., Hauck, J., Le Quéré, C., Peters, G. P., Peters, W., Pongratz, J., Sitch, S., Canadell, J. G., Ciais, P., Jackson, R. B., Alin, S. R., Anthoni, P., Bates, N. R., Becker, M., Bellouin, N., Bopp, L., Chau, T. T. T., Chevallier, F., Chini, L. P., Cronin, M., Currie, K. I., Decharme, B., Djeutchouang, L., Dou, X., Evans, W., Feely, R. A., Feng, L., Gasser, T., Gilfillan, D., Gkritzalis, T., Grassi, G., Gregor, L., Gruber, N., Gürses, Ö., Harris, I., Houghton, R. A., Hurtt, G. C., Iida, Y., Ilyina, T., Luijkx, I. T., Jain, A. K., Jones, S. D., Kato, E., Kennedy, D., Klein Goldewijk, K., Knauer, J., Korsbakken, J. I., Körtzinger, A., Landschützer, P., Lauvset, S. K., Lefèvre, N., Lienert, S., Liu, J., Marland, G., McGuire, P. C., Melton, J. R., Munro, D. R., Nabel, J. E. M. S., Nakaoka, S. I., Niwa, Y., Ono, T., Pierrot, D., Poulter, B.,

Rehder, G., Resplandy, L., Robertson, E., Rödenbeck, C., Rosan, T. M., Schwinger, J., Schwingshackl, C., Séférian, R., Sutton, A. J., Sweeney, C., Tanhua, T., Tans, P. P., Tian, H., Tilbrook, B., Tubiello, F., van der Werf, G., Vuichard, N., Wada, C., Wanninkhof, R., Watson, A., Willis, D., Wiltshire, A. J., Yuan, W., Yue, C., Yue, X., Zaehle, S. and Zeng, J. (2022). Global Carbon Budget 2021. *Earth System Science Data* 14: 1917–2005.

Garnaut, R. (2019). *Superpower: Australia's Low Carbon Opportunity.* La Trobe Univeristy Press, Melbourne, Australia.

Garnaut, R. (2021). *Rest: Restoring Australia After the Pandemic Recession.* La Trobe University Press, Melbourne, Australia.

Garnett, E. (2021). Meat eating is a big climate issue – but isn't getting the attention it deserve. *The Conversation*, 10 November 2021. Accessed at: https://theconversation.com/meat-eating-is-a-big-climate-issue-but-isnt-getting-the-attention-it-deserves-170855.

Global Carbon Capture and Storage Institute (2021). *Global Status of CCS Report 2021.* Accessed at: https://www.globalccsinstitute.com/resources/global-status-report/.

Gorman, A. (2021). *The Hill We Climb.* Chatto & Windus, London, UK.

Hardin, G. (1968). The Tragedy of the Commons. *Science* 162 (3859): 1243–1248.

Hawken, P. (2021). *Regeneration: Ending the Climate Crisis in One Generation.* Penguin Random House, New York, USA.

Heede, R. (2014). Tracing anthropogenic carbon dioxide and methane emissions to fossil fuel and cement producers, 1854–2010. *Climatic Change* 122 (1): 229–241.

Hertsgaard, M. and Pope, K. (2019). The media are complacent while the world burns. *Columbia Journalism Review*, 22 April 2019. Accessed at: https://www.cjr.org/special_report/climate-change-media.php.

Hickman, C., Marks, E., Pihkala, P., Clayton, S., Lewandowski, E., Mayall, E., Wray, B., Mellor, C. and van Susteren, L. (2021). Young people's voices on climate anxiety, government betrayal and moral injury: a global phenomenon. *The Lancet*: http://dx.doi.org/10.2139/ssrn.3918955.

High Ambition Coalition for Nature and People (2021). Why 30x30? Accessed at: https://www.hacfornatureandpeople.org.

International Energy Agency (2021a). *Net Zero by 2050: A Roadmap for the Global Energy Sector.* Accessed at: https://www.iea.org/reports/net-zero-by-2050.

International Energy Agency (2021b). *Renewables 2021.* Accessed at: https://www.iea.org/reports/renewables-2021.

International Energy Agency (2021c). *World Energy Outlook 2021.* Accessed at: https://www.iea.org/reports/world-energy-outlook-2021.

International Monetary Fund (2021). *Working Paper WP/21/236: Still Not Getting Energy Prices Right: A Global and Country Update of Fossil Fuel Subsidies.* Accessed at: https://www.imf.org/en/Publications/WP/Issues/2021/09/23/Still-Not-Getting-Energy-Prices-Right-A-Global-and-Country-Update-of-Fossil-Fuel-Subsidies-466004.

International Renewable Energy Agency (2021a). *Renewable Energy and Jobs: Annual Review 2021.* Accessed at: https://www.irena.org/-/media/Files/IRENA/Agency/Publication/2021/Oct/IRENA_RE_Jobs_2021.pdf.

International Renewable Energy Agency (2021b). *The Renewable Energy Transition in Africa: Powering Access, Resilience and Prosperity.* Accessed at: https://www.irena.org/-/media/Files/IRENA/Agency/Publication/2021/March/Renewable_Energy_Transition_Africa_2021.pdf.

IPCC (2019). *Climate Change and Land: An IPCC Special Report on Climate Change, Desertification, Land Degradation, Sustainable Land Management, Food Security, and Greenhouse Gas Fluxes in Terrestrial Ecosystems.* Shukla, P. R., Skea, J., Calvo Buendia, E., Masson-Delmotte, V., Pörtner, H. O., Roberts, D. C., Zhai, P., Slade, R., van Diemen, R., Ferrat, M., Haughey, E., Luz, S., Neogi, S., Pathak, M., Petzold, J., Portugal Pereira, J., Vyas, P., Huntley, E., Kissick, K., and Malley, J. Cambridge University Press, Cambridge, UK. Accessed at: https://www.ipcc.ch/srccl/.

IPCC (2022). Technical Summary. In *Climate Change 2022: Mitigation of Climate Change, the Working Group III contribution to the Sixth Assessment Report of the Intergovernmental Panel on Climate Change.* Masson-Delmotte, V., Zhai, P., Pirani, A., Connors, S. L., Péan, C., Berger, S., Caud, N., Chen, Y., Goldfarb, L., Gomis, M. I., Huang, M., Leitzell, K., Lonnoy, E., Matthews, J. B. R., Maycock, T. K., Waterfield, T., Yelekçi, O., Yu, R. and Zhou, B. (eds). Accessed at: https://www.ipcc.ch/report/ar6/wg3/.

IPCC, Canadell, J. G., Monteiro, P. M.S., Costa, M. H., Cotrim da Cunha, L., Cox, P. M., Eliseev, A. V., Henson, S., Ishii, M., Jaccard, S., Koven, C., Lohila, A., Patra, P. K., Piao, S., Rogelj, J., Syampungani, S., Zaehle, S., and Zickfeld, K. (2021). Chapter 5: Global Carbon and other Biogeochemical Cycles and Feedbacks Supplementary Material. In *Climate Change 2021: The Physical Science Basis. Contribution of Working Group I to the Sixth Assessment Report of the Intergovernmental Panel on Climate Change.* Masson-Delmotte, V., Zhai, P., Pirani, A., Connors, S. L., Péan, C., Berger, S., Caud, N., Chen, Y., Goldfarb, L., Gomis, M. I., Huang, M., Leitzell, K., Lonnoy, E., Matthews, J. B. R., Maycock, T. K., Waterfield, T., Yelekçi, O., Yu, R. and Zhou, B. (eds). Accessed at: https://www.ipcc.ch/report/ar6/wg1/.

Jackson, F. (2021). A 'Paris moment' for forests as world leaders commit to $19 billion in deforestation action. Accessed at: https://www.forbes.com/sites/feliciajackson/2021/11/02/world-leaders-commit-to-action-on-deforestation-with-19-billion-on-the-table/?sh=20dd6866743c.

Johnson, A. E. and Wilkinson, K. (2020). *All We Can Save: Truth, Courage and Solutions for the Climate Crisis*. One World, New York, USA.

Joy, M. (2010). *Why We Love Dogs, Eat Pigs and Wear Cows*. Conari Press, San Francisco, USA.

Julie's Bicycle (2021a). Culture: The Missing Link to Climate Action. Accessed at: https://juliesbicycle.com/news/the-british-council-executive-report/.

Julie's Bicycle (2021b). Julie's Bicycle: Creative Climate Action. Accessed at: https://juliesbicycle.com.

Kiehl, J. T. (2016). *Facing Climate Change: An Integrated Path to the Future*, Columbia University Press, New York, USA.

Klein Salamon, M. and Gage, M. (2020). *Facing the Climate Emergency: How to Transform Yourself with Climate Truth*. New Society Publishers, Gabriola Island, Canada.

Lamb, W. F., Wiedmann, T., Pongratz, J., Andrew, R., Crippa, M., Olivier, J. G. J., Wiedenhofer, D., Mattioli, G., Khourdajie, A. A., House, J., Pachauri, S., Figueroa, M., Saheb, Y., Slade, R., Hubacek, K., Sun, L., Ribeiro, S. K., Khennas, S., de la Rue du Can, S., Chapungu, L., Davis, S. J., Bashmakov, I., Dai, H., Dhakal, S., Tan, X., Geng, Y., Gu, B. and Minx, J. (2021). A review of trends and drivers of greenhouse gas emissions by sector from 1990 to 2018. *Environmental Research Letters* 16 (7): 073005.

Li, Y., Kalnay, E., Motesharrei, S., Rivas, J., Kucharski, F., Kirk-Davidoff, D., Bach, E. and Zeng, N. (2018). Climate model shows large-scale wind and solar farms in the Sahara increase rain and vegetation. *Science* 361 (6406): 1019-1022.

Low Carbon Power (2021). Ranking of Countries and Territories by Low-Carbon Electricity. Accessed at: https://lowcarbonpower.org/ranking.

Luntz, R. (2002). *The Environment: A Cleaner, Safer, Healthier America*. Accessed at: https://www.motherjones.com/files/LuntzResearch_environment.pdf.

Macy, J. and Johnstone, C. (2012). *Active Hope: How to Face the Mess We're in Without Going Crazy*. Finch Press, Sydney, Australia.

Mann, M. E. (2021). *The New Climate War: The Fight to Take Back Our Planet*. Public Affairs, New York, USA.

Maxmen, A. (2022). Wuhan market was epicentre of pandemic's start, studies suggest. *Nature* 603: 15-16.

McGreal, C. (2021). Big oil and gas kept a dirty secret for decades. Now they may pay the price. *The Guardian*, 30 June 2021. Accessed at: https://www.theguardian.com/environment/2021/jun/30/climate-crimes-oil-and-gas-environment.

McKibben, B. (2005). What the warming world needs now is art, sweet art. *Grist*, 22 April 2005. Accessed at: https://grist.org/article/mckibben-imagine/.

Meinshausen, M., Lewis, J., McGlade, C., Gütschow, J., Nicholls, Z., Burdon, R., Cozzi, L. and Hackmann, B. (2022). Realization of Paris Agreement pledges may limit warming just below 2°C. *Nature* 604 (7905): 304-309.

Milman, O. (2021). Meat accounts for nearly 60% of all greenhouse gases from food production, study finds. *The Guardian*, 14 September 2021. Accessed at: https://www.theguardian.com/environment/2021/sep/13/meat-greenhouses-gases-food-production-study.

Monbiot, G. (2021a). After the failure of COP26, there's only one last hope for our survival. *The Guardian*, 15 November 2021. Accessed at: https://www.theguardian.com/commentisfree/2021/nov/14/cop26-last-hope-survival-climate-civil-disobedience.

Monbiot, G. (2021b). Capitalism is killing the planet – it's time to stop buying into our own destruction. *The Guardian*, 30 October 2021. Accessed at: https://www.theguardian.com/environment/2021/oct/30/capitalism-is-killing-the-planet-its-time-to-stop-buying-into-our-own-destruction.

Muir, C., Wehner, K. and Newell, J. (2020). *Living with the Anthropocene: Love, Loss and Hope in the Face of Environmental Crisis*. NewSouth Publishing, Sydney, Australia.

New York Declaration on Forests (2021). *New York Declaration on Forests*. Accessed at: https://forestdeclaration.org.

Nunn, P. (2018). *The Edge of Memory: Ancient Stories, Oral Traditions and the Post-glacial World*. Bloomsbury Sigma, London, UK.

Oreskes, N., Conway, E., Karoly, D. J., Gergis, J., Neu, U. and Pfister, C. (2018). The denial of global warming. In *The Palgrave Handbook of Climate History*, pp. 149–171. White, S., Pfister, C. and Mauelshagen, F. London, Palgrave Macmillan, UK.

Poore, J. and Nemecek, T. (2018). Reducing food's environmental impacts through producers and consumers. *Science* 360 (6392): 987–992.

Rae, I. (2012). Saving the ozone layer: why the Montreal Protocol worked. *The Conversation*, 10 September 2021. Accessed at: https://theconversation.com/saving-the-ozone-layer-why-the-montreal-protocol-worked-9249.

Solnit, R. (2009). *A Paradise Built in Hell: The Extraordinary Communities That Arise in Disaster*. Viking, New York, USA.

Solnit, R. (2016). *Hope in the Dark: Untold Histories, Wild Possibilities*. Nation Books, New York, USA.

Solnit, R. (2021). Ten ways to confront the climate crisis without losing hope. *The Guardian*, 18 November 2021. Accessed at: https://www.theguardian.com/environment/2021/nov/18/ten-ways-confront-climate-crisis-without-losing-hope-rebecca-solnit-reconstruction-after-covid.

Taylor, M. and Watts, J. (2019). Revealed: the 20 firms behind a third of all carbon emissions. *The Guardian*, 9 October 2019. Accessed at: https://www.theguardian.com/environment/2019/oct/09/revealed-20-firms-third-carbon-emissions.

Teske, S., Nagrath, K., Morris, T. and Dooley, K. (2019). Renewable Energy Resource Assessment. *Achieving the Paris Climate Agreement Goals: Global and Regional 100%*

Renewable Energy Scenarios with Non-energy GHG Pathways for +1.5°C and +2°C, pp. 161–173. *Teske*, S. (ed.). Springer International Publishing, Cham, Switzerland.

Teske, S. and Niklas, S. (2021). *Fossil Fuel Exit Strategy: An Orderly Wind Down of Coal, Oil and Gas to Meet the Paris Agreement*. University of Technology, Sydney, Australia. Accessed at: https://fossilfueltreaty.org/exit-strategy.

Totally Renewable Yackandandah (2021). Totally Renewable Yackandandah. Accessed at: https://totallyrenewableyack.org.au.

United Nations Secretary-General (2022). Secretary-General's Video Message to the Press Conference Launch of IPCC Report. Accessed at: https://www.un.org/sg/en/node/262102.

United States Conference of Mayors (2017). Mayors strongly oppose withdrawal from Paris Climate Accord. The United States Conference of Mayors. Accessed at: https://www.usmayors.org/2017/06/01/mayors-strongly-oppose-withdrawal-from-paris-climate-accord/.

USA Spending (2021). The Federal Response to COVID-19. Accessed at: https://www.usaspending.gov/disaster/covid-19?publicLaw=all.

Victoria, M., Haegel, N., Peters, I. M., Sinton, R., Jäger-Waldau, A., del Cañizo, C., Breyer, C., Stocks, M., Blakers, A., Kaizuka, I., Komoto, K. and Smets, A. (2021). Solar photovoltaics is ready to power a sustainable future. *Joule* 5 (5): 1041–1056.

Vorrath, S. (2022). South Australia sets smashing new renewables record in final days of 2021. Accessed at: https://reneweconomy.com.au/south-australia-winds-up-2021-with-smashing-new-renewables-record/.

Waldron, A. (2021). *Protecting 30% of the Planet for Nature: Costs, Benefits and Economic Implications. Working Paper Analysing the Economic Implications of the Proposed 30% Target for Areal Protection in the Draft Post-2020 Global Biodiversity Framework*. Accessed at: https://www.conservation.cam.ac.uk/files/waldron_report_30_by_30_publish.pdf.

Wall Limmerer, R. (2015). *Braiding Sweetgrass: Indigenous Wisdom, Scientific Knowledge and the Teachings of Plants*. Milkweed Editions, Minneapolis, USA.

Weintrobe, S. (2012). *Engaging with Climate Change: Psychoanalytical and Inter-disciplinary Perspectives*. Routledge, London, UK.

Winkelmann, R., Donges, J. F., Smith, E. K., Milkoreit, M., Eder, C., Heitzig, J., Katsanidou, A., Wiedermann, M., Wunderling, N. and Lenton, T. M. (2022). Social tipping processes towards climate action: a conceptual framework. *Ecological Economics* 192: 107242.

Winton, T. (2022). Big Daddy Gas. *The Saturday Paper*, 5 March 2022. Accessed at: https://www.thesaturdaypaper.com.au/opinion/topic/2022/03/05/big-daddy-gas/164639880013448#hrd.

Wiseman, J. (2021). *Hope and Courage in the Climate Crisis: Wisdom and Action in the Long Emergency*. Palgrave Macmillan, Cham, Switzerland.

World Wildlife Fund (2020). *The Living Planet Report 2020*. Accessed at: https://livingplanet.panda.org.

Worobey, M. (2021). Dissecting the early COVID-19 cases in Wuhan. *Science* 374 (6572): 1202–1204.

Xu, X., Sharma, P., Shu, S., Lin, T.-S., Ciais, P., Tubiello, F. N., Smith, P., Campbell, N. and Jain, A. K. (2021). Global greenhouse gas emissions from animal-based foods are twice those of plant-based foods. *Nature Food* 2 (9): 724–732.

Permissions

Index